STUDY GUIDE

STUDY GUIDE

Terence J. Bazzett

State University of New York at Geneseo

to accompany

BRYAN KOLB

IAN Q. WHISHAW

AN INTRODUCTION TO BRAIN AND BEHAVIOR, Fourth Edition

WORTH PUBLISHERS

Study Guide
by Terence J. Bazzett
to accompany
Kolb/Whishaw: **An Introduction to Brain and Behavior, Fourth Edition**

ISBN 10: 1-4641-0990-7
ISBN 13: 978-1-4641-0990-4

Printed in the United States of America

First printing 2013

Worth Publishers
41 Madison Avenue
New York, NY 10010
www.worthpublishers.com

Contents

To the Student

Worried about doing well in this course? Relax. By opening this study guide you've taken an important step toward improving your grade. Now take the next step and spend a minute or two reading through the information below.

I've been teaching brain and behavior courses for over 25 years, and in that time I've observed the way my successful students prepare for quizzes, exams, and presentations. I've tried to incorporate that knowledge into the design of this study guide. My goal was to create a relatively simple tool that could enhance a student's performance without requiring an excessive amount of time or effort.

Each chapter of this study guide contains the following elements. For each item, I've offered a few suggestions for its effective use.

Summary: After you've read the chapter and listened to the lectures, read the summary quickly. It's okay to skim through this summary, as long as you're comfortable with your understanding of the key terms. If any of the terms look unfamiliar, you may wish to revisit your text to make sure you have a clear sense of how the term fits into the chapter.

Key Terms and *Key Names:* Take special note of key terms used by your professor in lectures and put stars by them in your study guide. If you're not completely clear on the meaning, look it up in the text and write a brief description next to the term. Some professors stress key names, whereas others may not refer to them at all. If your professor does not emphasize these names, do not labor to memorize them. Instead focus on terms. As a way to study for an exam try selecting a relatively large number of terms listed in this guide and stressed by your professor. Take a stack of 3- × 5-inch cards and randomly choose terms, writing about five on each card until you have a stack of cards containing terms that you think will appear on the upcoming exam. Get together with a few classmates and try a game that follows the rules of the word game Taboo. In general, this is how you play: One student selects a card and then tries to get his or her teammate to say the word by describing associated terms, concepts, and so on. Keep in mind that the person viewing the card can't use any form of the word to be guessed (no rhyming words either; these are "taboo"). Teammates then switch so that both get a chance to read and guess words. To determine a winning team, keep track of the number of correct terms and the amount of time to guess them. The most words in the shortest amount of time wins. This may sound corny, but if the alternative is sitting home alone trying to memorize terms, you may find this game of "bioboo" (as my students call it) less tedious, even a little enjoyable.

Practice Test: Don't panic if you don't get them all correct! This guide is very general and takes information from throughout each chapter without bias. Your professor will likely be much more specific about some topics, and may skip others completely. Missing a question that your professor has not covered and will not be using for a test is not cause for alarm. Essay questions are even more general than the multiple-choice questions, but I've tried to select topics that tend to interest a majority of professors. For each multiple-choice and essay question, I've given you a page number from your text where you will find the information needed to support the correct answer (though sometimes that information runs on for a page or two). Matching questions do not have page numbers listed for answers since many of them incorporate information from throughout the chapter. You can certainly look through and find the correct answers, but you can also just trust me on those. When I teach this course, diagram questions are among my favorites to use on tests. If your professor emphasizes drawings and diagrams, like I do, pay particular attention to this section of the Practice Test.

The Web: I've put in just a few of the many sites that correspond to some main topics in each chapter. Should your professor require a written project, these Web sites might jump-start your research. I've also tried to find some sites with particularly nice photos, videos, animations, and so on. One note of caution: Be sure to read the site addresses very carefully. Some have the www prefix, some do not. Some contain long strings of letters and numbers. Most are worth the trouble if you're interested in the topic. Unfortunately, it is possible that some of these sites will be defunct by the time you try them. Sorry, there's nothing we can do about that until the next edition of this guide is printed.

Crossword Puzzles: I love puzzles and I felt that there should be something particularly enjoyable in this guide, so I've included a relatively simple crossword puzzle for each chapter. These puzzles are actually "simple" only in terms of the number of words. Recalling these words (and correctly spelling them) can be very tricky. There is always the answer key if you get stuck. But before going to the answer key, try looking through the list of key terms. Most of the answers for these puzzles can be found in that list.

Good luck with your class.

Acknowledgment

I would like to thank Dr. Elaine Hull for her contributions to this and all of my works.

Good teachers answer students' questions.
Great teachers answer students' questions, and inspire them to ask more.

Thanks for the inspiration, Elaine.

CHAPTER

1 What Are the Origins of Brain and Behavior?

CHAPTER SUMMARY

Those who study neuroscience have long struggled to understand the relationship between *brain* and *behavior*, two inextricably linked yet vastly different entities. The brain may be described as an organ, as tissue or matter. Behavior, on the other hand, is far less tangible, described as an observable action but lacking in physical substance. The problem for researchers and philosophers alike has been to adequately explain how brain and behavior are related. This chapter sets the stage for the rest of the text by describing the brain, describing behavior, and then discussing theories of how the two are related.

The term *brain*, as used in the brain/behavior problem described above, generally refers to the entire *nervous system* which includes not only the brain but also the *spinal cord* and *peripheral nerves*, all of which play essential roles in the generation of behavior. The nervous system is an elaborate series of *nerve cells* that carry information to and from the brain and periphery via the spinal cord. Nerve cells that make up the brain and spinal cord collectively are referred to as the central nervous system (CNS), while all other nerve cells are considered to be part of the peripheral nervous system (PNS).

In the simplest terms for addressing the brain/behavior problem described above, it could be said that behavior consists of patterns in time. Such patterns may include both inherited and learned behaviors. Regarding the relationship between brain and behavior, a positive correlation has been noted between complexity of the nervous system and range of behaviors exhibited across species.

Theories of how brain and behavior are linked date back thousands of years, with early theories described as addressing the mind/body problem ("mind" describing the psyche or soul responsible for our behavior). *Aristotle* was among the first to philosophize about the relationship between mind and body, suggesting that human behavior is a product of the *psyche*, a nonmaterial entity that operates independent of material body organs, including the brain. This philosophy formed the basis for the theory of *mentalism*. Mentalism remained relatively unchallenged until the 1500s, when *René Descartes* proposed that the mind, though separate from the body, is somehow linked to physical organs, most notably the brain. This theory, proposing a nonphysical mind that was dependent upon the brain for receiving information and for controlling physical human behavior, formed the basis for the theory of *dualism*. More specifically, Descartes's theory of dualism placed the mind within the *pineal body* and put forth the hypothesis that pineal control over *ventricular fluid* was the basis of behavior. Dualism dominated the scientific approach to the study of brain and behavior until the mid-nineteenth century, when *Alfred Wallace* and *Charles Darwin* independently arrived at the same idea that all living things are related. Darwin advanced this theory by proposing the *theory of natural selection* as a means by which functional diversity can lead to an array of physical features as well as behaviors. *Gregor Mendel* lent support to Darwin's theory of natural selection by demonstrating that such functional diversity can evolve from *heritable factors* (which we now know are *genes*). Darwin's theory suggested that all species shared common behaviors as well as common organs (including the brain). His theories further suggested that the brain, and subsequently behavior, were built up over time through *evolution*. However, understanding that genetics alone cannot fully explain behavior, researchers typically consider the interplay between inherited genes (genotype), environment, and experience as they affect the expression of behavior. One contemporary school of thought, known as eliminative materialism, suggests behavior is a product of only physical brain matter and that intangible factors such as mind and consciousness do not need to be considered.

Although the nervous system (brain) is required for behavior, it is not essential for life. In fact, most living organisms lack a nervous system. Thus, nerve cells give animals the unique ability to create purposeful movement. Such movements range from those produced by the simplest *nerve net* (found in jellyfish), to a segmented nervous system (in flatworms), to a collection of neurons called *ganglia* (in clams and snails), to the most complex system of *chordates* containing a spinal cord and brain (in you). The size and complexity of the nervous system in chordates vary widely, but in general the evolutionarily most advanced species tend to have the largest *cerebral cortex* and *cerebellum* relative to the rest of the brain.

Present-day humans (*Homo sapiens*) are believed to represent a single descendant of numerous *hominid* (human-like animals) ancestors, including *Australopithecus africanus*. Among the early hominids that most closely resembled modern day humans were *Homo habilis* (handy human), *Homo erectus* (upright human), and *Neanderthals*. The relatively large brain of *Homo sapiens* is believed to have evolved through encephalization, which is a process of increasing the size and complexity of the cerebral cortex, which allows more complex behavior. Encephalization through evolution has resulted in a human brain that is larger in size (relative to body size) than any other animal. When considering that greater brain mass is required for increasingly sophisticated actions, this puts humans at the top of the scale for capacity to engage in complex behaviors.

Enlargement of the human brain is thought to be a response to changes in climate and food sources, and subsequently the need to develop strategies to adapt to these changes. For example, humans likely evolved into fruit gatherers, requiring more sensory and motor skills than grazing animals. Enlargement may also have been promoted by increased cerebral blood flow. Increased blood flow is associated with enhanced ability to cool the brain and presumably enhanced capacity for tissue growth and cell proliferation. *Neoteny* is the theory that humans represent a primate that expresses juvenile features of earlier ancestors, including a larger *cranium* (allowing for greater brain mass) relative to body size, and prolific nerve cell growth (as seen during development) over a longer period of time.

Within the human species, it is difficult to determine what brain factors (e.g., weight, volume), if any, correlate directly to intellectual capacity. Part of this problem results from an imprecise definition of what constitutes *intelligence*. In addition, brain size seems to correlate somewhat to body size, adding an additional extraneous variable to this task. Much of what we consider intelligent or complex behavior also seems to result from learning within our culture rather than ability inherent in brain cells.

KEY TERMS

The following is a list of important terms introduced in Chapter 1. Give the definition of each term in the space provided.

What Is the Brain?

Brain

Cerebrum

Hemispheres

Structure of the Nervous System

Neuron

Spinal cord

Central nervous system (CNS)

Peripheral nervous system (PNS)

Embodied language

Brain and Behavior

Psyche

Mind

Mentalism

Pineal body

Ventricles

Mind–body problem

Dualism

Materialism

Natural selection

Phenotype

Genotype

Epigenetics

Eliminative materialism

Evolution of Brain and Behavior

Common ancestor

Nerve net

Bilateral symmetry

Segmentation

Ganglia

Chordates

Cladogram

Human Evolution

Primate order

Hominids

Australopithecus

Homo habilis

Homo erectus

Neanderthals

Homo sapiens

Principle of proper mass

Encephalization quotient (EQ)

Radiator hypothesis

Neoteny

Species-typical behavior

Culture

KEY NAMES

The following is a list of important names introduced in Chapter 1. Explain the importance of each person in the space provided.

Aristotle

René Descartes

Alfred Russell Wallace

Charles Darwin

Gregor Mendel

Donald O. Hebb

Harry Jerison

PRACTICE TEST

Multiple-Choice Questions

Answer each of the following multiple-choice questions with the best possible answer based on information from your text.

1. According to your text, brain and behavior:
   A. differ greatly.
   B. are linked.
   C. have evolved together.
   D. are responsible for each other.
   E. All of the answers are correct.

2. Which of the following controls most of our conscious behavior?
   A. the brain stem
   B. the cerebrum
   C. the PNS
   D. the CNS
   E. the spinal cord

3. What is the term commonly given to the brain and spinal cord collectively?
   A. the nervous system
   B. the sensory pathway
   C. the motor pathway
   D. the peripheral nervous system
   E. the central nervous system

4. It has been proposed that the brain needs ongoing sensory and motor experiences to maintain intellectual activity. Supporting this theory is an experiment in which subjects who were deprived of all sensory input and motor output exhibited which of the following?
   A. significantly increased time spent sleeping
   B. inability to remember the experience
   C. a level of brain function similar to a coma
   D. an extremely unpleasant sensation, sometimes including hallucinations
   E. an extremely pleasant sensation, similar to that of meditation

5. The common crossbill inherits both the adaptive beak and the adaptive behavior for extracting seeds from pinecones. By comparison, the roof rat:
   A. inherits the adaptive behavior and adaptive crossed incisor teeth.
   B. inherits only the adaptive behavior to extract these seeds.
   C. learns the adaptive behavior from a parent to extract these seeds.
   D. learns the adaptive behavior from watching the crossbill.
   E. cannot extract seeds from these pinecones.

6. Aristotle was among the first to philosophize about the "mind," suggesting that it was:
   A. the same as the brain.
   B. the equivalent to what religious groups called the "soul."
   C. located in the heart.
   D. cooled by the blood.
   E. material in structure.

7. Following Aristotle were the theories of René Descartes that the brain and mind represented separate entities—one material and one nonmaterial, respectively. This theory formed the basis for which of the following philosophical positions?
   A. mentalism
   B. materialism
   C. dualism
   D. evolution
   E. All of the answers are correct.

8. Descartes was also among the first to suggest that the brain drove body function through mechanical means. What form of energy did he suggest was the basis for movement?
   A. thermal energy produced by heat from the brain
   B. electrical energy produced by individual cells in the brain
   C. hydraulic energy produced by movement of fluid from the brain
   D. chemical energy produced by chemicals in the brain
   E. None of the answers is correct.

9. The perspective of materialism disregards the need to consider which of the following?
   A. behavior
   B. the brain
   C. the mind
   D. evolution
   E. All of the answers are correct.

10. From the theories of Charles Darwin, and related research since his time, modern researchers believe which of the following regarding human emotions?
    A. Emotions are learned, not inherited.
    B. Human emotions differ across cultures.
    C. Emotions are not expressed by nonhuman animals.
    D. All of the answers are correct.
    E. None of the answers is correct.

11. Eliminative materialists might argue that consciousness can be improved from being a minimally conscious state (MCS), using which of the following?
    A. meditation
    B. behavior therapy
    C. deep brain stimulation
    D. hypnosis
    E. acupuncture

12. Although the first human-like brain probably evolved about 3 million to 4 million years ago, our modern human brain has "only" been around for about how many years?
    A. 1 million
    B. 500,000
    C. 200,000
    D. 50,000
    E. 10,000

13. The extremely simple nervous system found in older phyla, such as jellyfish, consists of which of the following?
    A. a brain with no spinal cord
    B. a spinal cord with no brain
    C. a series of connected ganglia
    D. a single ganglion
    E. a diffuse nerve net

14. The phyla that contains clams, snails, and octopuses represents the simplest nervous system that contains a collection of neurons termed a:
    A. ganglia.
    B. brain.
    C. spinal cord.
    D. central nervous system.
    E. None of the answers is correct.

15. The family of hominids, to which humans belong, also includes which of the following?
    A. gorillas
    B. chimpanzees
    C. neanderthals
    D. orangutans
    E. new world monkeys
    F. old world monkeys

16. The principle of proper mass, in general, states that:
    A. larger brains are associated with larger animals.
    B. larger animals tend to have more complex behaviors.
    C. larger animals tend to have less complex behaviors.
    D. larger brains are needed for increasingly complex behavior.
    E. brain size is not related to complexity of behavior.

17. Which of the following species from the human lineage has the largest encephalization quotient (EQ)?
    A. monkey
    B. chimpanzee
    C. *Homo sapiens*
    D. *Homo erectus*
    E. *Homo habilis*

18. Neoteny is one theory to explain why humans have developed such complex and large brains relative to other primates. Which of the following is true according to this theory?
    A. Adult humans have a greater capacity for neural development than do other adult primates.
    B. Adult humans have some physical features that resemble those of juveniles from other primates.
    C. Adult humans have some behavior patterns that resemble those of primate infants.
    D. It suggests a slowing of maturation compared to earlier ancestors.
    E. All of the answers are correct.

19. In attempting to correlate brain size to capacity for intellect, researchers measured the brain of Albert Einstein and found that it was:
    A. of average size.
    B. approximately 5% smaller than an average brain.
    C. approximately 5% larger than an average brain.
    D. approximately 10% larger than an average brain.
    E. approximately 20% larger than an average brain.

20. Art, language, mathematics and the like are elements of the brain's most remarkable achievment, which is the development of:
    A. long-term memory.
    B. elaborate motor skills.
    C. culture.
    D. cooperative living.
    E. primary sensory systems.

Short-Answer Questions

Answer each of the following questions with a brief but complete written answer based on information from your text.

1. Give three reasons, as cited in your text, why it is important to study both brain and behavior simultaneously.

2. Briefly define the central nervous system (CNS) and the peripheral nervous system (PNS).

3. Humans have the most complex nervous system of any animal, and thus the greatest capacity for learning new adaptive behaviors. However, humans still retain some inherited patterns of responding. Briefly explain why such basic behaviors are still present in humans.

4. Briefly describe how René Descartes' theory of dualism differed from Aristotle's earlier theory of mentalism.

5. Briefly describe the concept of natural selection, and give an example of how this process could lead to development of an increasingly complex nervous system.

6. Bilateral symmetry is a feature of both simple and complex nervous systems. Briefly describe bilateral symmetry and give an example of a very simple nervous system where bilateral symmetry is not found.

7. One theory of why humans have such large and complex brains is that this is an evolutionary adaptation to environment changes over the past several million years. Briefly explain why a brain might evolve greater complexity in response to changes in the animal's environment.

8. Briefly explain the "radiator hypothesis" of the evolution of human brain growth.

9. Briefly explain neotony and how this process might have contributed to the evolution of the human brain.

10. *Homo sapiens* today do not resemble *Homo sapiens* from 100,000 years ago in terms of intellect and behavior. Briefly explain what "culture" is and how it has influenced our behavior so dramatically over time.

Matching Questions

Complete each of the following matching questions based on information from your text.

1. Using taxonomy classifications, match the subgroup categories for humans to the appropriate classification.

A. Kingdom	___ Mammal
B. Phylum	___ Great apes
C. Class	___ Primates
D. Order	___ Animal
E. Family	___ Human
F. Genus	___ Chordates

2. Match each of the following historical names with their area of contribution to our current understanding of the relationship between the brain and behavior.

A. Mendel	___ Materialism
B. Aristotle	___ Dualism
C. Descartes	___ Genes
D. Darwin	___ Mentalism

3. Match each of the following names to their correct feature.

A. *Homo erectus*	___ Most recently evolved species
B. *Homo habilis*	___ Coexisted with early modern-day humans
C. Neanderthal	___ First to walk upright
D. *Homo sapiens*	___ Known for their use of tools

4. Rank order from 1 (lowest) to 5 (highest) of the encephalization quotients (EQs) of the following animals.

___ Rat
___ *Homo sapiens*
___ Elephant
___ Chimpanzee
___ Dolphin

The Web

Consider using the following Web sites for additional information on some of the topics from this chapter:

1. Philosophy of Mind: http://www.philosophyofmind.info/

2. Links to more philosophy of the mind sites: www.liv.ac.uk/~bdainton/mind2b.html

3. The Brain Injury Association Inc. (closed head injuries): www.biausa.org/

4. Charles Darwin: www.sc.edu/library/spcoll/nathist/darwin/darwin.html

5. Evolution of the brain: http://brainmuseum.org/evolution/index.html

CROSSWORD PUZZLE

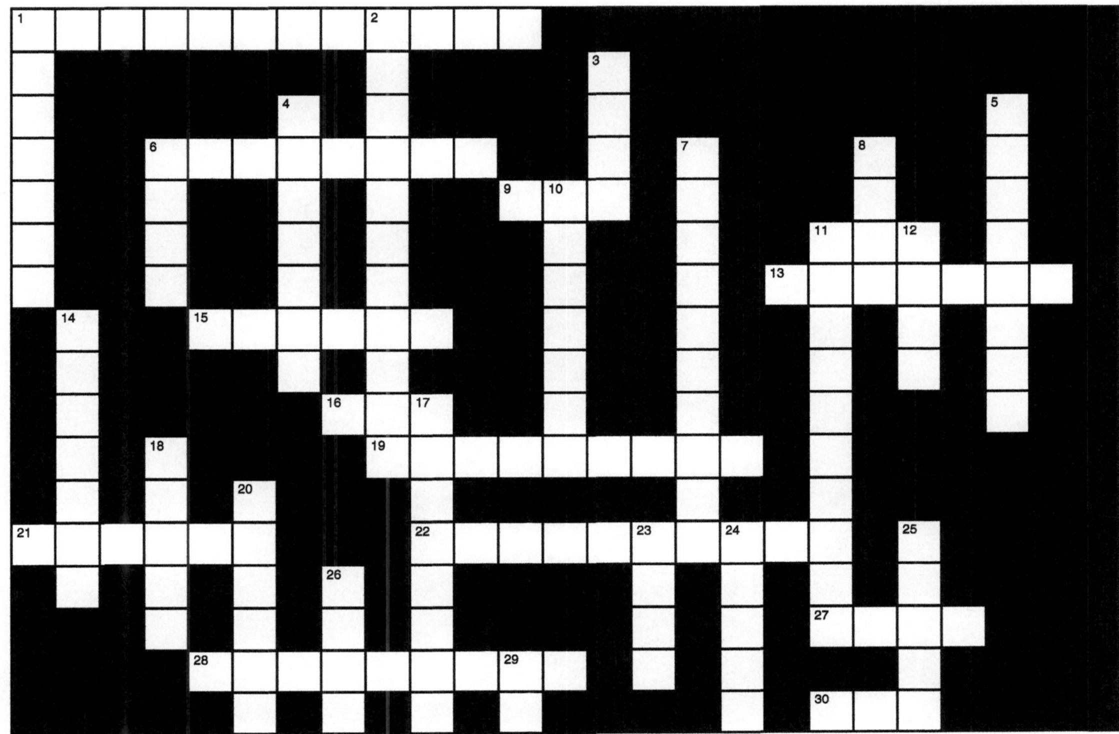

Across

1. For a while they co-existed with 4 down.
6. An action produced by the brain
9. Another for the order in 10 down
11. The brain and spinal cord; abbr.
13. This theory suggests your head looks a bit like that of a young chimp.
15. He did research with pea plants.
16. Simple nervous system, or a fishing aid
19. Nervous system organization as in a flatworm
21. Organs associated with minds
22. In nervous system terms, it's either central of_____.
27. Early philosophers equated it to the "soul"
28. Our phylum; named from the notochord
30. Not the CNS

Down

1. Darwin's _____ selection
2. 7 down plural . . . two per brain
3. Relatively small brained rodents
4. Modern human primate family; with 23 down
5. Human-like animals
6. Part of the crossbill that has evolved to a highly functional shape
7. Name for half a brain
8. Phenotype is the expression of a _____type.
10. The order for chimps and gorillas
11. The "little brain" in the back
12. The brain _____ connects the brain to 26 down.
14. Group behaviors, or a way to make yogurt
17. Brain lobe near your ear
18. Every culture expresses a positive emotional state with this behavior.
20. We call it "mind," Aristotle called it _____.
23. See 4 down.
24. Bilateral symmetry tells us the left hemisphere is a mirror image of the ____.
25. 15 across studied these in his plants
26. 28 across animals have this spinal component
29. Encephalization quotient, abbr.

CHAPTER

2 How Does the Nervous System Function?

CHAPTER SUMMARY

In Chapter 1 you learned that in simple terms, the function of the brain is to produce behavior. Chapter 2 expands on this concept to include, along with behavior, the brain functions of creating a *subjective reality* essential for carrying out complex tasks. Complex behaviors require a certain amount of neuroplasticity within the brain, thus allowing for changes in structure, organization, and function. Because it is so extensive, the nervous system is often subdivided into parts based on location and function. For example, the *central nervous system (CNS)* refers to the forebrain, brainstem, and spinal cord. The *peripheral nervous system (PNS)* can be further subdivided into the *somatic nervous system (SNS)* and the *autonomic nervous system (ANS)*. The SNS, made up of spinal and cranial nerves, sends and receives information associated with body movement and sensation. The ANS sends and receives information associated with activity of the internal organs, particularly during sympathetic (fight or flight) and parasympathetic (rest and digest) arousal. In addition, the direction in which information moves from one structure to another is generally either termed *afferent* (information coming into a structure) or *efferent* (information leaving a structure).

The language, or *nomenclature*, used to describe the brain is important because names given to brain structures may describe aspects of function, location, or physical features of those structures. This nomenclature is at times confusing, due in part to the fact that the brain has been studied for centuries by many cultures; as a result, contributions to the scientific literature span time and languages. In general, the relative location of structures is usually defined using the terms *medial* (middle), *lateral* (side), *dorsal* (top), *ventral* (bottom), *anterior* (front), and *posterior* (back). The terms *superior* and *inferior* are sometimes used to describe dorsal and ventral, respectively. Sometimes the terms *rostral* and *caudal* are used to describe anterior and posterior, respectively. To view internal structures, the brain is usually sectioned in one of three different orientations. Slicing sections from front to back produces coronal sections. Slicing sections from top to bottom produces horizontal sections. Slicing the brain from the side toward the center line produces sagittal sections.

Upon visual inspection, it becomes clear that the brain is wrapped in connective tissue. There are actually three layers of tissue, also known as *meninges*. A tough, fibrous outermost layer is called the *dura mater*. Beneath the dura mater are the *arachnoid layer* and the *pia mater*. These two layers are thinner and less durable than the dura mater and tend to follow all the contours of the brain's uneven surface. *Cerebral spinal fluid (CSF)* circulates between the arachnoid and pia mater layers, providing a cushion for the brain.

Gross observation of the brain reveals a *cerebrum* (divided evenly into two *hemispheres*) and a *cerebellum*. The outer layer of these structures appears very convoluted (wrinkled). As discussed in Chapter 1, this wrinkling allows more surface area to be compressed into the confines of the skull. The wrinkles also have names. *Gyri* are raised surfaces, whereas *sulci* are the indentations. Very deep sulci are called *fissures*. The area where the brain connects to the spinal cord is called the *brainstem*, and it is at this level that protruding cranial nerves can be seen. Finally, an elaborate system of blood vessels covers the entire brain.

When the brain is cut into sections, several gross features can be clearly identified. Some internal regions appear white because of a high density of myelinated nerve fibers, while other areas appear gray because of a high density of cell bodies. Areas of *white matter* thus represent regions associated with information transmission while areas of *gray matter* represent regions of information processing. Hollow, fluid-filled cavities called the *ventricles* can be seen in several areas. It is within these ventricles that CSF is produced. The *corpus callosum*, a large band of white matter connecting the two hemispheres, can also be clearly identified. Beyond these few large structures, it is difficult to identify brain structures with the naked eye. Inspection with a microscope, however, reveals a wide array of cell types and connections in the brain tissue. Neurons can be distinguished from glia by the presence of *axons* (nerve fibers) that connect neurons. Glial cells lack this distinctive feature.

The developing brain of mammals is less complex than that of the mature adult brain and is biologically similar in many ways to the brain of simpler, more primitive species such as amphibians. In other words, the elements of these more primitive nervous systems are present in the most complex systems, including ours. Shared features include bilateral symmetry and the spinal cord. The adult brain of a fish, an amphibian, or a reptile can also be roughly divided into three parts that correspond to parts of the human brain. The prosencephalon (front brain) of primitive animals develops into the cerebral cortex in humans, the mesencephalon (middle brain) develops into subcortical structures including the cerebral aqueducts, and the rhombencephalon (hindbrain) develops into the cerebellum and other brainstem structures.

The spinal cord is an integral part of the CNS. Although the brain controls most movement, some very basic motor patterns (e.g., knee jerk reflex) are controlled by the spinal cord. Bridging the brain and spinal cord is the brainstem. The *brainstem* may be subdivided into several distinct regions, each containing numerous structures. The *hindbrain* region is composed of deep internal structures that sit atop the spinal cord and are responsible for motor, sensory, and integrative functions. These structures are the *medulla*, *reticular formation*, *cerebellum*, and *pons*. The medulla controls vital functions such as breathing, heart rate, and blood pressure. The reticular formation is responsible for waking and sleeping. The cerebellum is responsible for rapid complex movements. The pons is made up of fibers connecting the cerebellum to the rest of the brainstem and acting as a bridge for information exchange.

The *midbrain* region consists of the *superior* and *inferior colliculi* (collectively called the *tectum*) and the *tegmentum*. The superior colliculus processes information about location of visual stimuli. The inferior colliculus has a similar function for auditory information. Likely, the function of the tectum is to localize all types of sensory input. The tegmentum is responsible for some aspects of movement. The tegmentum contains the *substantia nigra*, a structure particularly important for initiating movement. This region also contains the periaqueductal gray area identified by a high concentration of opiate receptor–containing cell bodies important in pain modulation.

The *diencephalon* primarily consists of two structures called the *thalamus* and the *hypothalamus*. Each of these two structures is made up of over 20 smaller nuclei, with each of these nuclei likely influencing different behavioral functions. For example, the hypothalamus controls temperature regulation, feeding, sleeping, emotions, and sexual behaviors (to name a few). The hypothalamus also directly controls the *pituitary gland*. The thalamus receives a large amount of incoming sensory information and likely acts as a way station, or switchboard, for processing and directing this information to appropriate brain regions.

The forebrain consists of three principal structures: the *cortex*, the *basal ganglia*, and the *limbic* system. The cortex includes the outer convoluted structure seen upon gross inspection. In the human brain, the cortex comprises approximately 80 percent of the total volume of the brain. It is believed that the cortex is involved in the highest level processing of sensory information, including thinking or cognition. Neurons from the cortex innervate nearly every structure in the brain, and likely influence nearly every aspect of behavior. In general the cortex is divided into four regions or *lobes*. These lobes (mentioned in Chapter 1) are the *frontal* (anterior-most region), *occipital* (posterior-most region), *parietal* (dorsal region between frontal and occipital) and *temporal* (ventral region between frontal and occipital). The cortex can also be divided into five distinct layers, each responsible for a different aspect of information processing. The five primary structures that make up the basal ganglia are the *caudate nucleus*, *putamen*, *globus pallidus*, *substantia nigra*, and *subthalamic nucleus*. Collectively, these structures play a vital role in control of movement. Parkinson's disease and Tourette's syndrome both have a basis in malfunction in this system and both result in disrupted movement. The limbic system includes the *amygdala*, *hippocampus*, and the *cingulate cortex*. The limbic system controls many aspects of emotional and sexual behavior, as well as some aspects of learning and memory formation.

The SNS includes 12 pairs of cranial nerves responsible for both sensory input and motor output associated with head and face function. Also part of the SNS, the *spinal nervous system* is composed of the *spinal cord*, encased within the bony *vertebrae* that make up the backbone. The spinal cord is segmented, with nerves running to and from particular body regions (or dermatomes). These nerve fibers carrying information from the body enter the spinal cord through the dorsal region (termed the *dorsal root*). Fibers sending information to the body exit the spinal cord from the *ventral root*. When sliced into sections, a distinctive region of cell bodies within the spinal cord appears gray, while regions of nerve fibers entering and exiting the cord appear white. In general the spinal cord conveys information to and from the brain, but some simple reflexive behaviors function independently of brain activity and are mediated by cells within the spinal cord.

The *internal nervous system* is commonly termed the *autonomic nervous system*, which controls internal organs and glands. The autonomic nervous system is actually composed of two opposing systems termed the *sympathetic* and the *parasympathetic nervous systems*. When activated the sympathetic nervous system arouses the body, stimulating increased heart rate and blood pressure while decreasing digestive function. In contrast, parasympathetic nervous system arousal causes an increase in digestion and a decrease in heart rate and blood pressure.

Ten principles can be applied to nervous system function:

1) The fundamental function of the nervous system is to produce behaviors in response to environmental factors interpreted by the brain.

2) The nervous system exhibits neuroplasticity, which is required for learning, memory, and survival.

3) A peculiar feature of our nervous system is that sensory and motor processes associated with the left side of our body are controlled by the right hemisphere of our brain, and vice versa.

4) As animals evolved to engage in more complex behaviors, new brain structures were added to existing structures. This evolutionary process has resulted in multiple levels of brain functioning, with the most complex organisms utilizing the greatest number of levels.

5) Regarding the two hemispheres of the brain, it has been noted that primary control of a few behaviors is localized to a single hemisphere. Language in humans is considered a primary function of the left hemisphere, with far less influence coming from the right. Conversely, spatial concepts are generally processed in the right hemisphere.

6) The nervous system is composed of systems that process information both in parallel and in hierarchical fashion. However, because these systems are so intimately linked it is difficult to establish independent features of each. Despite this difficulty, the approach taken by neuroscientists is that each aspect of behavior may be localized to some region (or regions) of the nervous system.

7) Most neurons are likely designated as processing either sensory or motor signals. Although this is obvious in neurons receiving input from the environment or those sending signals to muscles, even within intermediate regions receiving and transmitting a very large number of signals (like the cortex), researchers have designated motor and sensory regions and layers.

8) Sensory input likely originated as a means of controlling motor output. A second but highly useful function of sensory input is object recognition.

9) While specific aspects of complex behavior are often localized, such behaviors often integrate input from multiple regions that may be widely distributed.

10) It is important to realize that behaviors may result not only from excitation of a particular area, but also from active inhibition of an area. We intuitively understand that damage to a structure can result in loss of behavior; but also, some regions actively stop behaviors from occurring. In Huntington's disease, for example, degeneration of parts of the basal ganglia results in abnormal motor behaviors that include flaying limb movements. Similarly, Tourette's syndrome causes degeneration of a part of the basal ganglia that results in facial tics and uncontrollable vocalizations, including swear words in some cases. Such diseases represent loss of a structure normally exerting inhibitory influence over expression of behaviors.

KEY TERMS

The following is a list of important terms introduced in Chapter 2. Give the definition of each term in the space provided.

Overview of Function and Structure

Phenotypic plasticity

Plastic

Neuroplasticity

Peripheral nervous system (PNS)

Central nervous system (CNS)

Somatic nervous system (SNS)

Autonomic nervous system (ANS)

Afferent

Efferent

Nomenclature

Dorsal

Ventral

Medial

Lateral

Anterior

Posterior

Rostral

Caudal

Superior

Inferior

Coronal section

Sagittal section

Horizontal section

The Brain's Surface Features

Meninges

Dura mater

Arachnoid layer

Pia mater

Cerebrospinal fluid (CSF)

Cerebrum

Cerebral Cortex

Cerebellum

Temporal lobe

Frontal lobe

Parietal lobe

Occipital lobe

Gyri

Sulci

Fissures

Hemispheres

Brainstem

Stroke

The Brain's Internal Features

Coronal section

Sagittal section

Horizontal section

White matter

Gray matter

Ventricles

Corpus callosum

Subcortical regions

Neuron

Glial cell

Nuclei

Cells and Fibers

Axons

Nerve

Tract

Evolutionary Development of the Nervous System

Prosencephalon

Mesencephalon

Rhombencephalon

Telencephalon

Diencephalon

Metencephalon

Myelencephalon

Spinal cord

Brainstem

Forebrain

Hindbrain

Midbrain

Reticular formation

Pons

Medulla

Tectum

Superior colliculus

Inferior colliculus

Tegmentum

Substantia nigra

Periaqueductal gray matter

Hypothalamus

Thalamus

Pituitary gland

Limbic system

Basal ganglia

Neocortex

Limbic cortex

Frontal lobe

Parietal lobe

Temporal lobe

Occipital lobe

Cytoarchitectonic maps

Top-down processing

Caudate nucleus

Putamen

Globus pallidus

Tourette's syndrome

Amygdala

Hippocampus

Cingulate cortex

Cingulate gyrus

Olfactory system

Olfactory bulb

Pyriform cortex

Somatic

Somatic Nervous System

Dermatome

Cranial nerves

Dorsal root

Ventral root

Law of Bell and Magendie

Automatic Nervous System

Autonomic nervous system

Sympathetic system

Parasympathetic system

10 Principles of Nervous System Function

Contralateral

Ipsilateral

Excitation

Inhibition

Parallel system

Hierarchical system

PRACTICE TEST

Multiple-Choice Questions

Answer each of the following multiple-choice questions with the best possible answer based on information from your text.

1. Which of the following best describes the term *afferent* as it refers to neural information?
   A. active
   B. inactive
   C. coming into a region of the nervous system
   D. going out of a region of the nervous system
   E. coming before a behavior

2. The meninges is a three-layered structure encasing the brain and composed primarily of connective tissue. Which of the following is the Latin name for the tough outermost layer?
   A. pia mater
   B. dura mater
   C. arachnoid layer
   D. meninges externalis
   E. None of the answers is correct.

3. Based on the name, where would you expect to find the superior colliculus, relative to the inferior colliculus?
   A. Superior colliculus would be located dorsal from the inferior colliculus.
   B. Superior colliculus would be located ventral from the inferior colliculus.
   C. Superior colliculus would be located medial from the inferior colliculus.
   D. Superior colliculus would be located lateral from the inferior colliculus.
   E. Location cannot be determined from the name.

4. Both the cerebrum and the cerebellum have very wrinkled surfaces. Which of the following terms is *not* used to describe these convolutions?
   A. gyri
   B. sulci
   C. denti
   D. fissures
   E. All of the answers are correct.

5. Which of the following is produced within the ventricles in the brain?
   A. neural impulses
   B. electrical impulses
   C. cerebral spinal fluid
   D. blood plasma
   E. oxygen

6. A collection of nerve fibers (axons) found outside of the CNS is simply referred to as "nerves." However, when such a fiber bundle is located within the CNS, it is generally referred to as a:
   A. ganglion.
   B. nucleus.
   C. tract.
   D. structure.
   E. cord.

7. The diencephalon is:
   A. also referred to as the "end brain."
   B. found in the spinal cord.
   C. not a part of the mammalian brain.
   D. contains a structure called the hypothalamus.
   E. All of the answers are correct.

8. Which of the following is true of the cerebellum?
   A. It is well developed in animals that move slowly and steadily.
   B. It is located ventral from the pons and medulla.
   C. It is one of the smallest structures in the brain.
   D. It is shaped like a carrot.
   E. None of the answers is correct.

9. If the medulla were severely damaged, which of the following would most likely occur?
   A. The animal would show disrupted movement.
   B. The animal would show disrupted memory.
   C. The animal would show disrupted waking and sleep patterns.
   D. The animal would show disrupted emotional behavior patterns.
   E. The animal would likely die.

10. Located in the tectum are the superior and inferior colliculi. The superior colliculus receives a great deal of sensory input from the visual system. It is not surprising that the adjacent inferior colliculus:
   A. also receives a great deal of sensory input from the visual system.
   B. receives a great deal of sensory input from the auditory system.
   C. controls most eye movements.
   D. controls visual aspects of dreams.
   E. controls pupil dilation.

11. The substantia nigra is a nucleus located within the tegmentum. Which of the following best describes the function of the substantia nigra?
   A. It is involved in memory.
   B. It is involved in sexual behavior.
   C. It is involved in the initiation of movement.
   D. It is involved in vital respiratory and cardiovascular functions.
   E. It is involved in sleep behavior.

12. Which of the following is the most distinctive functional feature of the thalamus?
   A. It controls all aspects of movement.
   B. It acts as a gateway for all sensory input being sent to the cortex.
   C. It has inhibitory control over nearly every other structure in the brain.
   D. It processes input for all sensory systems except the visual system.
   E. It produces cerebral spinal fluid.

13. The neocortex is unique in several features. Which of the following describes a unique feature of the neocortex?
   A. It is involved in mental processes such as perception and planning.
   B. It comprises the majority of brain volume in humans.
   C. It can be subdivided into several distinct layers, each with a unique function.
   D. It can be subdivided into several regions called lobes, each with a unique function.
   E. All of the answers are correct.

14. Gilles de la Tourette's syndrome is believed to be primarily a result of basal ganglia dysfunction. Which of the following might be a symptom of Tourette's syndrome?
   A. difficulty breathing
   B. loss of balance
   C. extreme weight loss
   D. extreme weight gain
   E. involuntary cursing

15. The limbic system includes the amygdala, cingulate cortex, and hippocampus. When would you be most likely to utilize your hippocampus?
   A. when watching television
   B. when listening to the radio
   C. when studying for an exam
   D. when walking on a path through the woods
   E. when sleeping

16. How many cranial nerves have been identified as protruding from the brainstem?
    A. 12
    B. 12 pairs
    C. 21
    D. 21 pairs
    E. The total number of cranial nerves has not yet been determined.

17. Which of the following is *not* one of the five groups of vertebrae found in humans?
    A. basal
    B. thoracic
    C. lumbar
    D. sacral
    E. cervical
    F. coccygeal

18. Which of the following is *not* true of the sympathetic nervous system?
    A. Activation increases heart rate.
    B. Activation increases digestion.
    C. It acts to arouse the body.
    D. It works in opposition to the parasympathetic nervous system.
    E. Initiation of this system arises from the spinal cord.

19. Damage to a brain structure in your right hemisphere that controls motor function would likely lead to disruption of movement in which of the following?
    A. the right side of your body
    B. the left side of your body
    C. both sides of your body
    D. neither side of your body, since motor control is primarily a function of the spinal cord
    E. There is no way to determine what side of the body will be affected based on what side of the brain is damaged.

20. Which of the following most accurately describes the structure of the cortex as it relates to sensory and motor functions?
    A. There are certain regions of the cortex designated for sensory functions.
    B. There are certain layers of the cortex designated for sensory functions.
    C. There are certain regions of the cortex designated for motor functions.
    D. There are certain layers of the cortex designated for motor functions.
    E. All of the answers are correct.

Short-Answer Questions

Answer each of the following questions with a brief but complete written answer based on information from your text.

1. There seems to be a rather diverse nomenclature (vocabulary) used to describe the structures of the brain. Regions may have names of Latin, Greek, or English origin. Some are described with numbers, others with letters. Briefly explain why such diversity is seen in the nomenclature.

2. Cerebral spinal fluid is a clear fluid that surrounds the brain and flows through the ventricular system and into the spinal cord. Name at least three possible functions of this fluid.

3. The term *pons* literally means “bridge.” Where is this structure located and why is it referred to as a bridge?

4. Although the hypothalamus comprises less than 1 percent of the brain’s weight, it is intimately involved in many aspects of behavior. List at least four behaviors that are influenced by the hypothalamus.

5. The neocortex has developed as a structure with six layers of gray matter. Briefly describe the physical and functional differences that make these layers distinct.

6. List two different diseases that are characterized by degeneration of the basal ganglia. What is the common behavioral manifestation seen in these two diseases?

7. List as many of the 12 cranial nerves as you can recall. Include the function mediated by each nerve.

8. Briefly describe the law of Bell and Magendie.

9. Although many behaviors are controlled by both hemispheres, several functions are lateralized; that is, they are controlled by a single hemisphere in the brain. Language is a good example, being controlled primarily by the left hemisphere. Explain briefly why language might be better controlled by one hemisphere rather than two.

10. Damage to neurons providing excitatory input to a region that produces a behavior will result in a reduction or loss of that behavior. Briefly explain the effect of damage to neurons providing inhibitory input to that same structure. Use a specific behavior as an example.

Matching Questions

Complete each of the following matching questions based on information from your text.

1. Match the following neuroanatomy terms with their more common counterparts.

	__ Anterior
A. Toward the front	__ Posterior
B. Toward the back	__ Ventral
C. Toward the middle	__ Medial
D. Toward the side	__ Superior
E. Toward the top	__ Inferior
F. Toward the bottom	__ Caudal
	__ Rostral

2. Match the following nervous system to the best descriptor.

A. Brain and spinal cord	__ Peripheral nervous system (PNS)
B. Everything outside of the CNS	__ Sympathetic nervous system
C. A subdivision of the autonomic nervous system	__ Central nervous system (CNS)
D. Also called the "autonomic nervous system"	__ Internal nervous system

3. Match the following structure to the best descriptor.

A. Fibers connecting the two hemispheres	__ Hypothalamus
B. Produces and transports CSF	__ Ventricles
C. Used for complex coordinated movements	__ Neocortex
D. It mediates pituitary gland function	__ Cerebellum
E. It can be divided into 6 layers of gray matter	__ Corpus callosum

4. List the following regions of the brain from most dorsal (1) to most ventral (4).

___ Cortex
___ Midbrain
___ Diencephalon
___ Hindbrain

5. Match the following types of brain damage with the regions they are most likely to affect.

A. Rabies	___ Basal ganglia
B. Bell's palsy	___ May affect many regions of the brain
C. Parkinson's disease	___ Pia mater and arachnoid layer
D. Meningitis	___ Limbic system
E. Stroke	___ Facial nerves

Diagrams

1. Identify the cerebrum, the cerebellum, the brainstem, and the olfactory bulb on the diagram of the cat brain below.

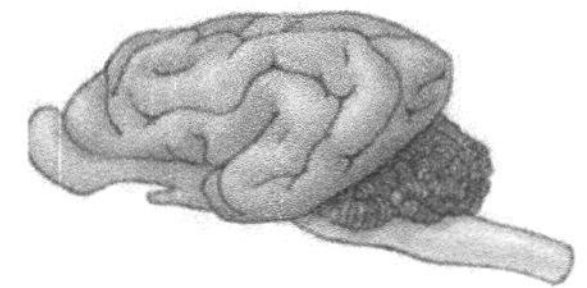

2. Imagine the diagram below represents a coronal section of the hypothalamus. You have been told that damage to the ventromedial hypothalamus will affect eating behavior in rats. Indicate (in general) where the ventromedial hypothalamus is located.

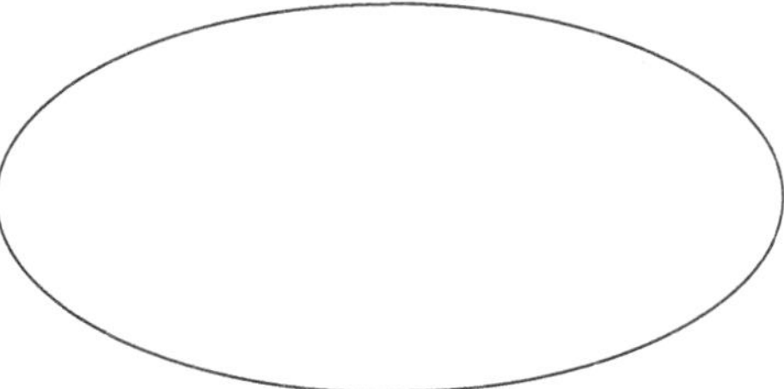

3. Identify the four lobes of the cerebral cortex in the diagram of a human brain below.

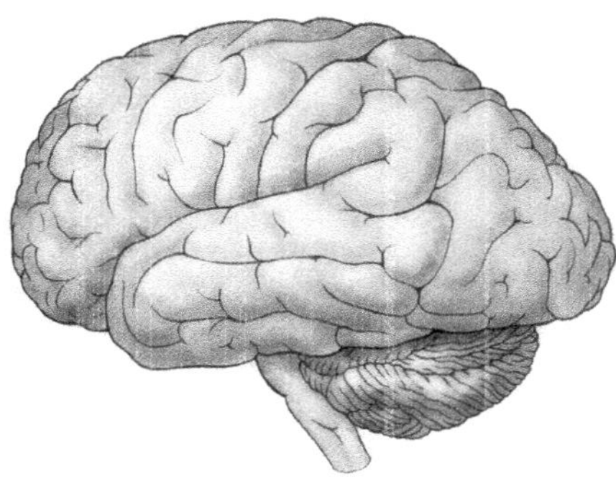

4. Appropriately label the following structures of the hindbrain and midbrain on the diagrams below: pons, medulla, cerebellum, superior colliculus, inferior colliculus.

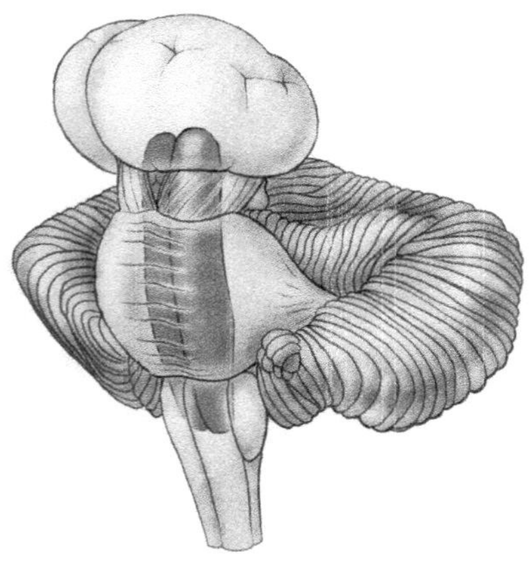

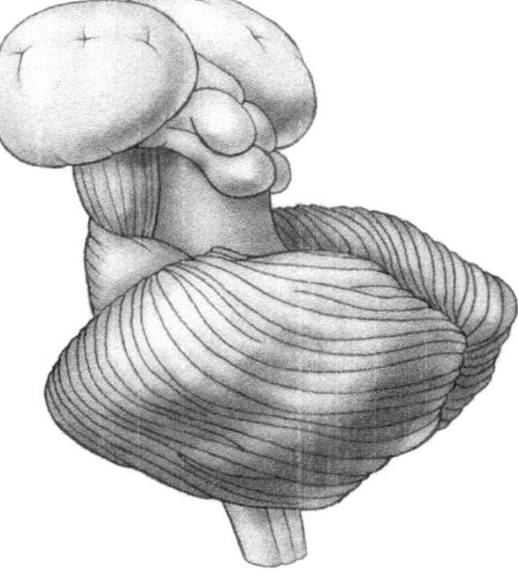

5. Appropriately label the following structures of the forebrain on the diagram below: cerebral cortex, basal ganglia, hippocampus, amygdala.

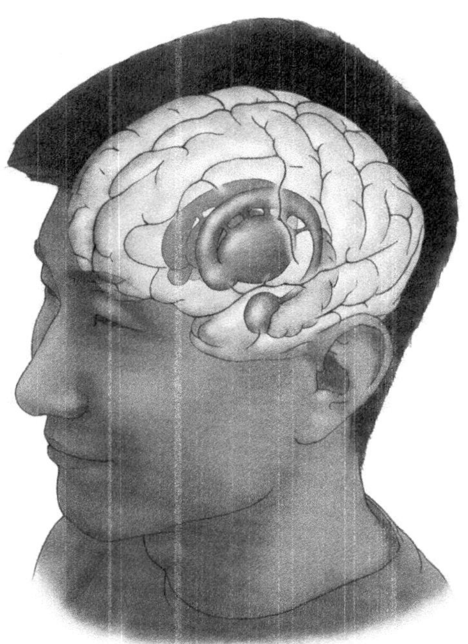

6. Below is a cross section of the spinal cord. Identify the following: ventral root, dorsal root, gray matter, white matter.

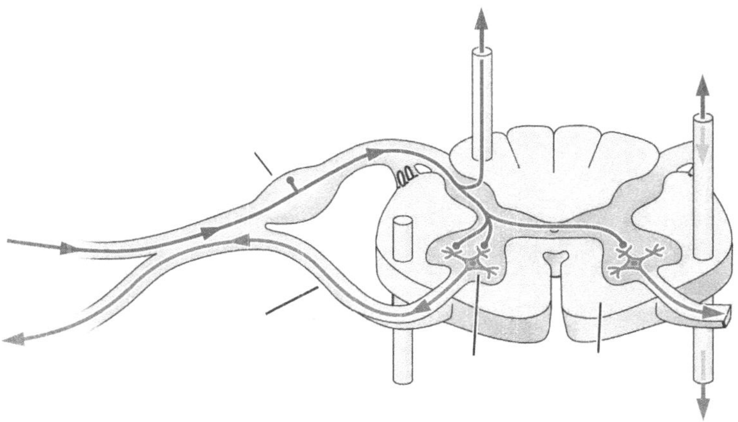

7. Draw a simple diagram of a spinal cord below. Identify the approximate areas associated with the following five segments: cervical, thoracic, lumbar, sacral, and coccygeal.

8. On the diagram below, draw lines to indicate each of the following: sensory input from the right hand to the appropriate hemisphere, motor output from the right hemisphere to the appropriate hand, olfactory input from the left nostril to the appropriate hemisphere.

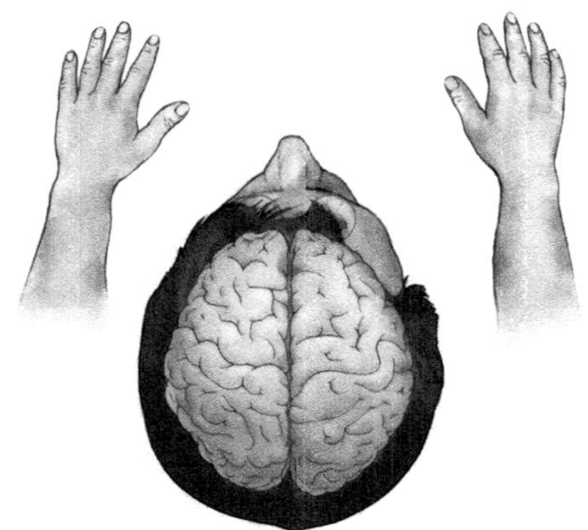

The Web

Consider using the following Web sites for additional information on some of the topics from this chapter:

1. Interactive Brain site: www.newscientist.com/movie/brain-interactive
2. Neuroanatomy tutorial: www.gwc.maricopa.edu/class/bio201/brain/1neuro.htm
3. Gross brain structure and function: www.sci.uidaho.edu/med532/start.htm
4. Interactive brain anatomy: www.brainexplorer.org
5. Cranial nerve anatomy: http://faculty.washington.edu/chudler/cranial.html

CROSSWORD PUZZLE

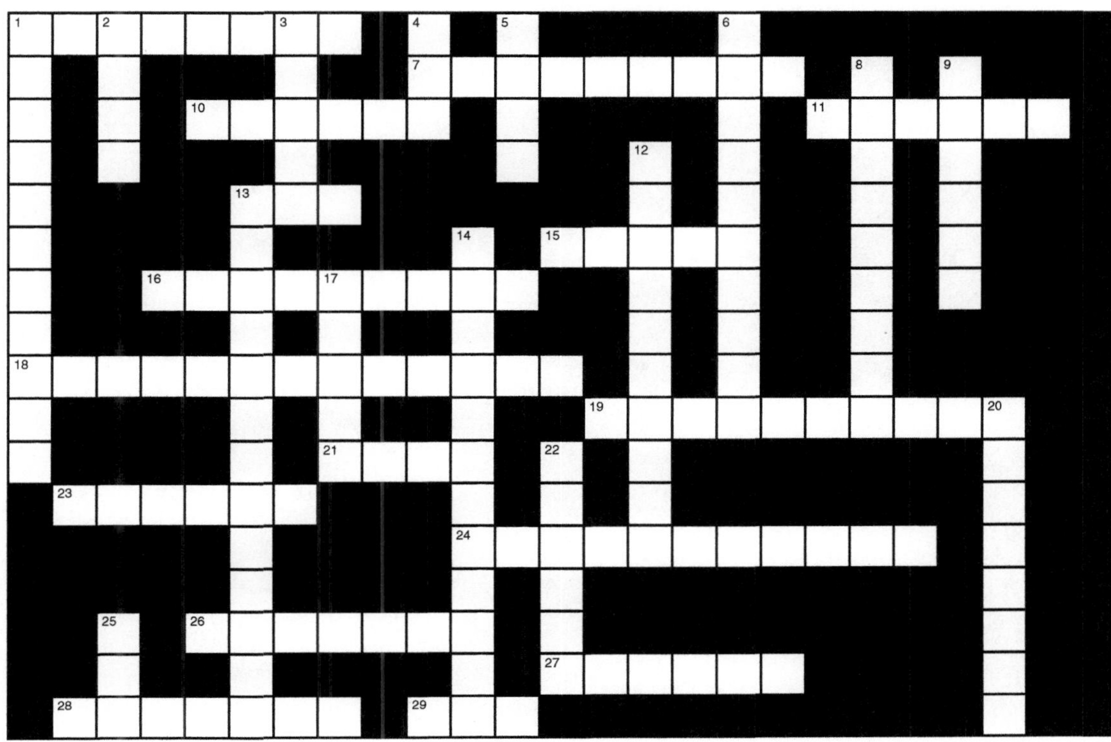

Across

1. A front-to-back cut through the brain
7. The "new" outer layer of the brain
10. Mediates movement; _____ pallidus
11. Common term for 10 down
13. Liquid in 28 across; abbr.
15. Term for shallow fissures in 10 down
16. Region below the midbrain
18. The first two words of 24 down; abbr.
19. Mediates movement; _____ nigra
21. Temporal or occipital
23. Opposite of ventral
24. 11 down connects them
26. The most anterior 20 across
27. Band of white fibers in the middle of the brain; _____ callosum
28. See 18 across
29. All neurons outside of 26 across; abbr.

Down

1. Term meaning "below" 24 across
2. Name for small bumps on 10 down
3. 11 across is made up mostly of these
4. Brain and spinal cord; abbr.
5. Dorsal or ventral of the spine
6. You have 2 laterals, a 3rd and a 4th
8. Above the midbrain
9. Common term for disrupted blood flow to brain
12. Vision to the superior, auditory to the inferior
13. Term for "the other side" of the brain
14. Contains hypothalamus and thalamus
17. A collection of nuclei controlling movement with 21 down
20. Structure controlling emotions in 5 across
22. System containing hippocampus and 14 down
25. Meninges layer; _____ mater

CHAPTER

3 What Are the Functional Units of the Nervous System?

CHAPTER SUMMARY

Throughout history researchers have struggled to define the vast functions of neurons. While the basic structures of soma, dendrite, and axon have long been established, it is at times difficult to imagine the functional complexity of even a single brain cell. For example, single neurons are responsible not only for relatively simple tasks of acquiring and passing on information, but also for tasks as complex as encoding memories. Making the job of determining how neurons accomplish such tasks even more difficult is the fact that neurons exhibit a great deal of plasticity over time. In other words, neurons are constantly changing in structure and, consequently, in function throughout life.

The basic description of a neuron is that it is a specialized cell capable of receiving, integrating, assessing, and ultimately sending information. Some neurons are specialized to receive and pass information from the environment to our brain (*sensory neurons*). Other neurons are specialized to receive and pass information from our brain to our muscles (*motor neurons*). The majority of neurons, however, are cells responsible for processing information within the brain. These cells are generally categorized as *interneurons* and comprise the major difference in brain mass between those animals with small brains and those with large brains.

Although some neurons utilize electrical signals for communication at their *synapses*, the vast majority of neurons communicate using *chemical messengers*. In addition to communicating with other cells, neurons may also receive *feedback* from their own signals in determining if the signal has been adequately sent, or if continued chemical release is required. The initiation of neural signaling is ultimately determined by "weighing" excitatory and inhibitory input. A cell may only initiate firing when excitatory signals outweigh inhibitory signals. The versatility inherent in weighing excitatory and inhibitory signals from multiple interneuron inputs results in enormous possibilities for behavior.

Glial cells are often described as "support cells" within the nervous system. Unlike neurons, glial cells have not yet been found to establish functional connections for cellular communication. They do, however, engage in numerous activities essential for neural communication and proper brain function. Among these tasks, *ependymal glial cells* produce and secrete cerebral spinal fluid used to mobilize nutrients, eliminate waste products, cushion and possibly cool the brain. *Astroglia*, sometimes called astrocytes, provide a structural framework for neurons and secrete substances needed for maintaining health of neurons and healing of injured neurons. Astroglia also comprise the *blood–brain barrier*, binding blood-vessel cells tightly together to inhibit unwanted substances from entering the brain from the bloodstream. These cells also may cause blood vessels to dilate in areas where greater blood flow is needed for increased neural activity. If neurons are injured, astroglia also form a type of scar tissue that is beneficial to the healing process. *Microglia* have a primary function of phagocytosis, engulfing and removing debris left by dead tissue. *Oligodendroglia* provide myelin used to insulate axons and speed neurocommunication in the brain and spinal cord. *Schwann cells* have a similar role in the peripheral nervous system, with the additional benefit of being able to guide regrowth of axons that have degenerated as a result of injury. *Multiple sclerosis* is a disease in which myelin degenerates, resulting in slower and sometimes confused (short-circuited) neural communication.

Electrically charged atoms called *ions* comprise the active *natural elements* found in intracellular and extracellular fluid. These elements may also be found bound together as components of larger *molecules*. For example, the element sodium (Na^+) may bind to the element chloride (Cl^-) to form a salt (NaCl) molecule. Understanding the role of ions, elements, and molecules required for neural function is one very basic step toward understanding neural control of behavior.

The parts of the cell required for proper function are sometimes likened to work centers in a factory. The *cell membrane*, a *lipid bilayer* designed to keep substances out of the cell, is like an exterior factory wall. A similar membrane surrounding the *nucleus* (*nuclear membrane*), like an interior factory wall, keeps elements of the *intracellular fluid* (*cytosol*) from freely entering the nucleus. *Endoplasmic reticulum* is where *protein products* are assembled in the cell. *Golgi bodies* are where these protein products are packaged and readied for transport. *Microtubules* transport packaged products to different regions within the cell. *Mitochondria* power the cell and all of its functions. *Lysosomes* act as a cleaning and maintenance crew, arranging incoming supplies and disposing of waste products. The nucleus is where the blueprints are kept for all products produced by the cell. These blueprints are *genes* coded in the chemical structure of nuclear *chromosomes*. Copies of these *DNA* blueprints are transported from within the nucleus to intracellular *organelles* in the form of *mRNA*. It is from this mRNA that the organelles of our cell know what type of peptide chain or protein is to be assembled. In particular, the endoplasmic reticulum, which contains a high concentration of *ribosomes*, translates the mRNA and begins the production process. The number of possible combinations of peptide chains that can be manufactured from our 20 different available amino acids is nearly countless. Many of the proteins and polypeptide chains manufactured in the *soma* are transported by motor molecules to sites where they may be imbedded into the cell membrane or excreted from the cell through the process of *exocytosis*. Substances imbedded in the membrane can take the form of *receptors*, *gates*, *channels*, or *pumps*. Receptors respond to neurotransmitters. Gates open and close to allow ion diffusion through membrane channels. Pumps actively transport substances in and out of the cell.

Genes comprise chromosomes and supply the blueprints used to produce the proteins essential for all aspects of neural function. As such, genes ultimately contribute heavily to behavior. With tens of thousands of genes contributing to brain development, it is difficult to speculate on individual contributions. However, it is known that some genetic abnormalities can have severe consequences on behavior. Studying such genetic disorders can provide insight into contributions of some genes. Every normally developed individual has 23 pairs of chromosomes containing between 20,000–25,000 genes. About half of our genes code for proteins used in the development of the brain. One pair (*sex chromosomes*) determines primary development of sex characteristics. As chromosomes are paired, genes that make up those chromosomes are also paired into matching *alleles*. When like alleles are matched, the pair is termed *homozygous*. When alleles containing different instructions are matched, the pair is termed *heterozygous*. In the case of a heterozygous pair, the genetic instructions from one allele are generally expressed to a greater extent. This is termed a *dominant* allele, which usually results in greater influence over the *phenotype* (outward appearance) of the organism. *Recessive* phenotypes are generally expressed only when two recessive alleles comprise a homozygous pair. For example, *Tay-Sachs disease* results only when two recessive alleles are present in the individual. On the other hand, *Huntington's chorea* (which results from an abnormal dominant allele) is expressed in individuals with only one affected allele. Recessive and dominant features of these genetic abnormalities directly affect the mathematical probability that a person may inherit the disorder from parents who are carriers or affected individuals. *Down syndrome* is an example of a disorder that results from the addition of an entire chromosome to the genetic makeup of an individual. This disorder is also termed trisomy 21, referring to three 21st chromosomes where only two should exist. Each of these genetic diseases has a significant impact on development and behavior. As such, researchers have learned a great deal about the contributions of individual chromosomes, and even individual genes, to behavior. Technology has recently been developed whereby scientists can manipulate the genetic structure of rodents. With this ability, researchers have been able to develop more accurate models of genetic disorders. Perhaps more important, they are able to assess the general behavioral changes associated with genetic manipulation in an effort to more fully understand the functional role of these most basic contributors to human behavior. In addition, a new line of research has begun to investigate factors capable of modifying the underlying genetic mechanisms responsible for phenotypic expression. *Epigenetic factors* can modify the functioning of histones, DNA or RNA, ultimately producing variation in the output of identical DNA coding.

KEY TERMS

The following is a list of important terms introduced in Chapter 3. Give the definition of each term in the space provided.

Structure and Function of the Neuron

Soma

Axon

Axon hillock

Axon collaterals

Teleodendria

End foot/terminal button

Synapse

Dendrite

Dendritic spines

Types of Neurons

Bipolar neuron

Interneuron

Somatosensory neuron

Motor neuron

Stellate cell

Pyramidal cell

Purkinje cell

Excitation

Inhibition

Glial Cells

Ependymal cells

Cerebrospinal fluid

Hydrocephalus

Astroglia (astrocytes)

Blood–brain barrier

Microglia

Phagocytosis

Oligodendroglia

Myelin

Schwann cell

Multiple sclerosis

Internal Structure of the Cell

Organelles

Element

Atom

Neutron

Proton

Electron

Ion

Molecule

Polar molecule

Hydrogen bond

Parts of a Cell

Hydrophobic

Hydrophilic

Membrane

Phospholipid

Nucleus

Nuclear membrane

Endoplasmic reticulum (ER)

Golgi body

Microtubule

Microfilament

Mitochondria

Lysosome

Extracellular fluid

Intracellular fluid

The Nucleus

Chromosome

Gene

Deoxyribonucleic acid (DNA)

Nucleotide base

Adenine

Thymine

Guanine

Cytosine

Transcription

Messenger RNA (mRNA)

Protein Synthesis, Packaging, and Shipment

Ribosome

Translation

Codon

Transfer RNA (tRNA)

Amino group

Carboxyl group

Peptide bond

Polypeptide chain

Motor molecule

Exocytosis

Cell Membrane

Protein

Receptor

Channel

Gate

Pump

Genes, Cells, and Behavior

Genotype

The Human Genome Project

Sex chromosome

Autosome

Allele

Homozygous

Heterozygous

Wild-type allele

Mutation

Trait

Phenotype

Dominant allele

Recessive allele

Complete dominance

Incomplete dominance

Codominance

Mutation

Genetic Disorders and Genetic Engineering

Tay-Sachs disease

Huntington's chorea

Down syndrome

Cloning

Chimeric animal

Transgenic animal

Knockout technology

Phenotypic plasticity

Epigenetic mechanisms

KEY NAMES

The following is a list of important names introduced in Chapter 3. Explain the importance of each person in the space provided.

Camillo Golgi

Santiago Ramón y Cajal

Barbara Webb

Gregor Mendel

PRACTICE TEST

Multiple-Choice Questions

Answer each of the following multiple-choice questions with the best possible answer based on information from your text.

1. By our best estimates, the human nervous system contains around:
   A. one hundred thousand neurons.
   B. one million neurons.
   C. one hundred million neurons.
   D. one billion neurons.
   E. one hundred billion neurons.

2. Neurons show a surprising ability to change form and structure throughout life. This is particularly apparent in which of the following?
   A. ability of neurons to multiply
   B. ability of neurons to add additional nuclei
   C. neurons losing old, and producing new, connections
   D. neurons changing to glial cells
   E. None of the answers is correct.

3. Which of the following best describes the location of terminal buttons?
   A. touching the terminal buttons of other neurons
   B. close to, but not quite touching, the terminal buttons of other neurons
   C. touching the dendritic spines of other neurons
   D. close to, but not quite touching, the dendrite spines of other neurons
   E. The location of terminal buttons is highly variable and cannot be described in specific terms.

4. Which of the following would *not* be considered a type of neuron?
   A. ependymal cell
   B. pyramidal cell
   C. Purkinje cell
   D. interneuron
   E. bipolar neuron

5. Which of the following is *not* a likely function of cerebral spinal fluid secreted by ependymal cells?
   A. cushion the brain when the head is jarred
   B. eliminate waste products
   C. provide nutrients to the brain
   D. mediate electrical signals between neurons
   E. cool the brain

6. The blood–brain barrier is more of a concept than an actual structure. However, if pressed to describe the makeup of the blood–brain barrier, you could say it is composed primarily of:
   A. astroglia and proteins.
   B. astroglia and blood vessels.
   C. microglia and proteins.
   D. microglia and blood vessels.
   E. The blood–brain barrier contains all of the above.

7. The process by which oligodendroglia and Schwann cells form myelin sheaths around axons could best be described as which of the following?
   A. continuous growth of myelin originating from the soma to the terminal button
   B. continuous growth of myelin originating from the terminal button to the soma
   C. continuous growth of myelin originating from both the terminal button and the soma
   D. Intermittent segments of axon are wrapped by outgrowths from adjacent glial cells.
   E. All of the answers are correct.

8. In addition to forming an insulating myelin sheath, Schwann cells also provide which function for peripheral neurons following damage?
   A. act as scavengers to clean debris left by dead axons
   B. form scar tissue to seal the area around dead axons
   C. guide regrowth of axons after damage
   D. repel regrowth of axons by releasing an antigrowth agent
   E. All of the answers are correct.

9. An atom that becomes positively or negatively charged by gaining or losing an electron is called:
   A. an element.
   B. a trace element.
   C. a neutron.
   D. a proton.
   E. an ion.

10. Which of the following structures within the cell "packages" newly synthesized proteins in preparation for transporting them?
    A. Golgi bodies
    B. endoplasmic reticulum
    C. microtubules
    D. mitochondria
    E. lysosomes

11. The neuronal membrane can be described as a lipid bilayer formed by phospholipid molecules composed of a "head" and "tail." More specifically, the molecules form the membrane with:
    A. hydrophobic heads facing intracellular fluid and hydrophilic tails facing extracellular fluid.
    B. hydrophilic heads facing intracellular fluid and hydrophobic tails facing extracellular fluid.
    C. hydrophobic heads facing intra- and extracellular fluid, and hydrophilic tails facing each other.
    D. hydrophilic heads facing intra- and extracellular fluid, and hydrophobic tails facing each other.

12. The chromosomes within a cell nucleus contain which of the following?
    A. genes
    B. DNA
    C. nucleotide bases
    D. coded instruction for development of all proteins in the cell
    E. All of the answers are correct.

13. The primary function of mRNA is to:
    A. synthesize proteins.
    B. produce the code for protein synthesis.
    C. transport genetic code out of the nucleus.
    D. generate DNA.
    E. provide energy for the nucleus.

14. Regarding the relationship between a protein and a polypeptide chain, it could be said that:
    A. they are two terms to describe the same components.
    B. a protein is formed by a particular configuration of a polypeptide chain.
    C. a polypeptide chain is formed by a particular configuration of a protein.
    D. a polypeptide chain consists of many proteins.
    E. there is no general relationship between these two components.

15. Excretion of proteins through a cell membrane is generally accomplished through which of the following?
    A. via passive diffusion
    B. via mRNA
    C. via microtubules
    D. via motor molecules
    E. via exocytosis

16. One goal of the Human Genome Project is to catalog all of the human genes. How many genes are humans estimated to possess?
    A. over 1,000,000
    B. 500,000–1,000,000
    C. 100,000–500,000
    D. 50,000–100,000
    E. less than 50,000

17. Genetic literature often uses the term *wild type* to describe an allele being studied. In this case, "wild type" refers to which of the following?
    A. an allele that produces abnormal behavior
    B. an allele that produces exaggerated behavior
    C. an allele that does not occur in domesticated organisms
    D. a mutated form of a common allele
    E. a common allele from which mutations may occur

18. Which of the following could *not* be considered a possible description of a genetic mutation?
    A. a change in a single nucleotide base
    B. a change that produces a beneficial change in development of the organism
    C. a change that produces a disruptive change in development of the organism
    D. no more than one may occur in any single gene
    E. responsible for human hereditary disorders

19. Which of the following is *least* likely to be characteristic of a transgenic mouse developed to study Huntington's chorea?
    A. It would have an experimentally induced alteration to its normal genetic makeup.
    B. It could potentially to be used for testing therapeutic treatments for Huntington's chorea.
    C. It would exhibit abnormal movements.
    D. It would have abnormally low production of the protein huntingtin.
    E. It would have unusually extensive cell death in some brain regions.

20. Considering Mendelian genetics and human genetic disorders, which child is most likely to be affected by (express the phenotype of) an inherited disease?
    A. a child with one parent who is heterozygous for Tay-Sachs disease
    B. a child with one parent who is heterozygous for Huntington's chorea
    C. a child with two parents who are heterozygous for Tay-Sachs disease
    D. a child with two parents who are heterozygous for Huntington's chorea
    E. a child with one parent who is homozygous for Tay-Sachs disease

Short-Answer Questions

Answer each of the following questions with a brief but complete written answer based on information from your text.

1. The particular structure of a neuron is generally a good indicator of the primary function of that neuron. With this in mind, describe the structure of an interneuron and how that particular structure serves the function of this cell type. Contrast this description with that of a somatosensory neuron. Explain how the somatosensory neuron structure serves the function of this type of cell.

2. List the five different types of glial cells and note a primary function for each type.

3. One important function of astroglia is the ability to convey signals from neurons to blood vessels. What is this important signal that neurons send to blood vessels via astroglia, and when would you expect this message to be sent?

4. Define phagocytosis and identify which type of cell is most likely to engage in this activity.

5. Describe the symptoms of multiple sclerosis. Explain in terms of neuronal function the cause of these symptoms.

6. What is a molecule? Give an example of a molecule that is intimately involved in neural function.

7. Each strand of DNA possesses a variable sequence of four nucleotides. These four nucleotides always occur in two predictable pairings. Name the four nucleotides and identify how they are paired in strands of DNA.

8. Describe the general function of each of the following membrane proteins:

   Receptor:

   Channel:

   Gate:

   Pump:

9. What is the probability that a child born to two parents heterozygous for Tay-Sachs disease will be affected by this disorder? What is the probability that a child born to one parent heterozygous for Huntington chorea will be affected by this disorder? Explain your answer.

10. Describe both the genotype and the phenotype of a Down syndrome individual.

    Genotype:

    Phenotype:

Matching Questions

Complete each of the following matching questions based on information from your text.

1. Match the following characteristics to the MOST appropriate structure:

A. Vesicles	__ Soma
B. Receptor sites	__ Terminal
C. Nucleus	__ Axon
D. Myelin	__ Dendrite

2. Match the following cell types to their general description:

A. Motor	__ Single short axon and single short dendrite
B. Pyramidal	__ Dendrite connected directly to axon
C. Purkinje	__ Many dendrites extend directly from soma
D. Stellate	__ Long axon with two sets of dendrites that extend from soma
E. Somatosensory	__ Many dendritic branches form a fan shape
F. Bipolar	__ Extensive dendrites, large soma, long axon to muscle

3. Identify the following glial cells:

astrocyte (astroglia), microglia, oligodendroglia, Schwann cell, ependymal

_____________ _____________ _____________ _____________ _____________

4. Match the following characteristics to the MOST appropriate cell part or organelle:

A. Phospholipid bilayer	__ Cell membrane
B. Contains DNA	__ Endoplasmic reticulum
C. Assembles proteins	__ Nucleus
D. Packages proteins	__ Microtubules
E. Transports proteins	__ Golgi bodies
F. Provides cell energy	__ Lysosomes
G. Moves and stores waste	__ Mitochondria

5. Match the following disorders to the best descriptor regarding genetic contribution:

A. Huntington's chorea	__ Caused by recessive allele
B. Tay-Sachs disease	__ Caused by dominant allele
C. Down syndrome	__ No known genetic contribution
D. Multiple sclerosis	__ Caused by additional chromosome

Diagrams

1. Assume messenger output from the sensory neuron below is excitatory. Draw axons and axon collaterals from sensory neuron to interneurons, and from interneurons to motor neurons so that motor neuron A. is actively inhibited, motor neuron B. is unaffected, and motor neuron C. is actively excited. Use a + to indicate excitation and a – to indicate inhibition. Be sure all motor neurons receive interneuron input.

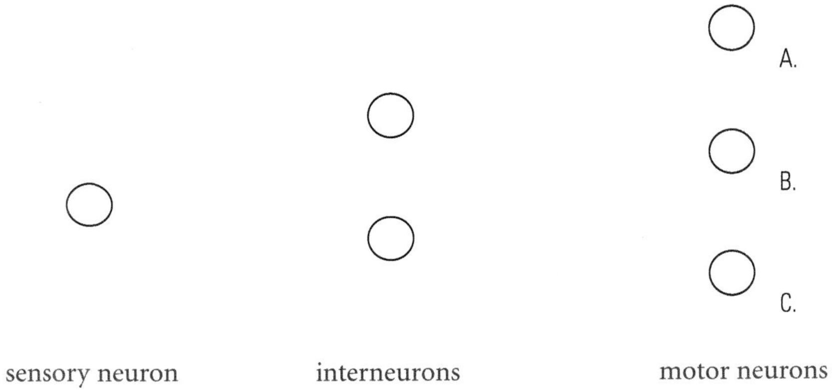

A.

B.

C.

sensory neuron interneurons motor neurons

2. Indicate the location of the following structures on the neuron below:

   cell body (soma), axon, dendrites, dendritic spines, axon hillock, axon collateral, teleodendria, end foot, terminal button, nucleus

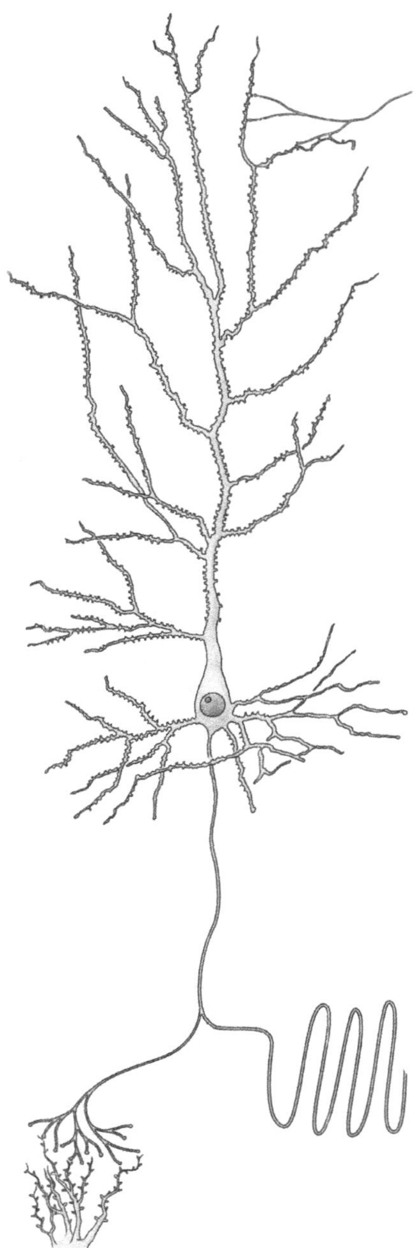

3. Identify the following cells and cell structures in the figure below:

   astrocyte, myelinated axon, blood-vessel cells, neuron

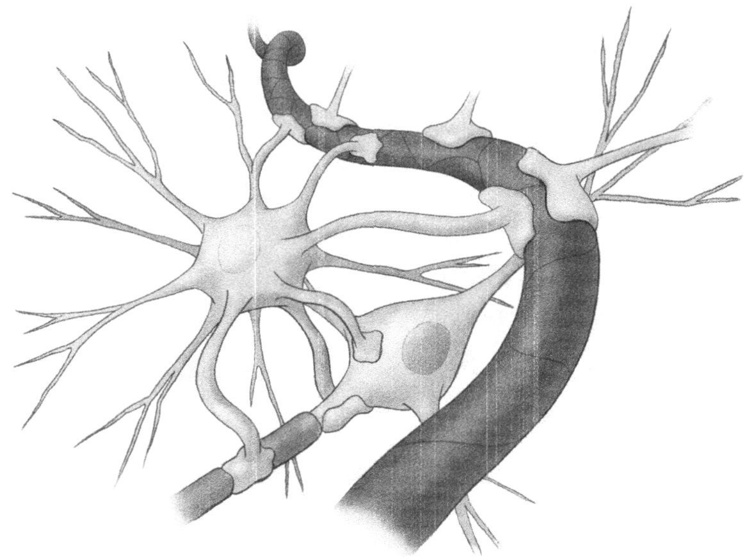

4. Identify the following organelles in the figure below:

   nucleus, endoplasmic reticulum, mitochondrion, microtubules, Golgi apparatus

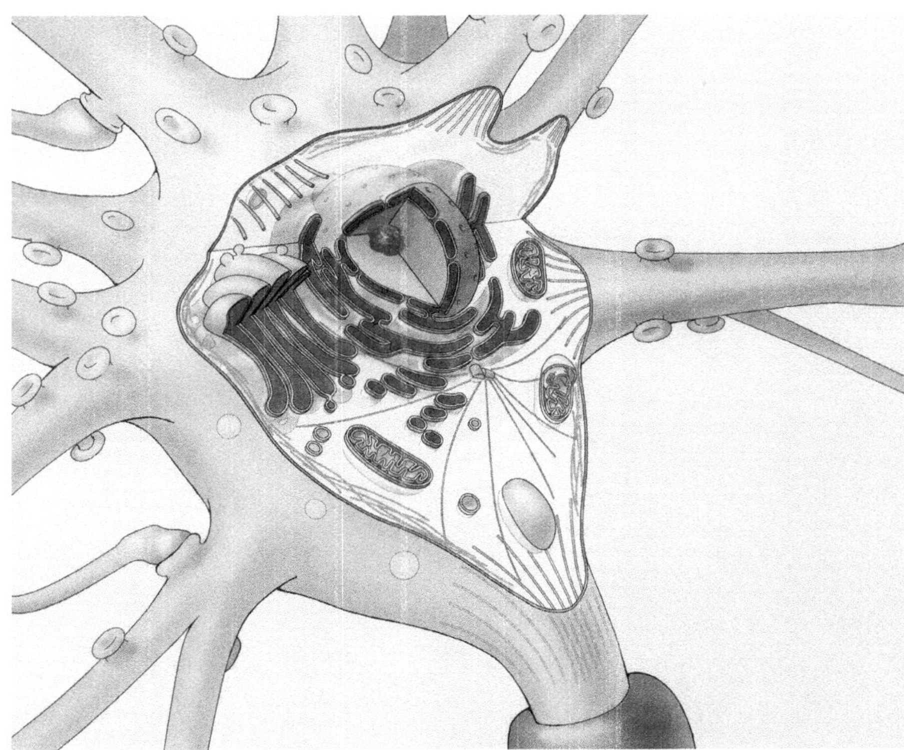

5. Assuming N indicates a chromosome with a normal gene, H indicates a chromosome with a Huntington's chorea gene, and T indicates a chromosome with a Tay-Sachs disease gene, determine the probability that offspring from the following pairs will express the phenotype of the disease.

   HN + NN   HN + HN   HH + NN   TN + NN   TN + TN   TT + NN

The Web

Consider using the following Web sites for additional information on some of the topics from this chapter:

1. National Tay-Sachs & Allied Diseases Association: www.ntsad.org
2. Internet Resources for Huntington's Disease: www.hdsa.org
3. National Multiple Sclerosis Society: www.nmss.org/
4. National Association for Down Syndrome (NADS): www.nads.org
5. Hereditary Disease Foundation: www.hdfoundation.org/

CROSSWORD PUZZLE

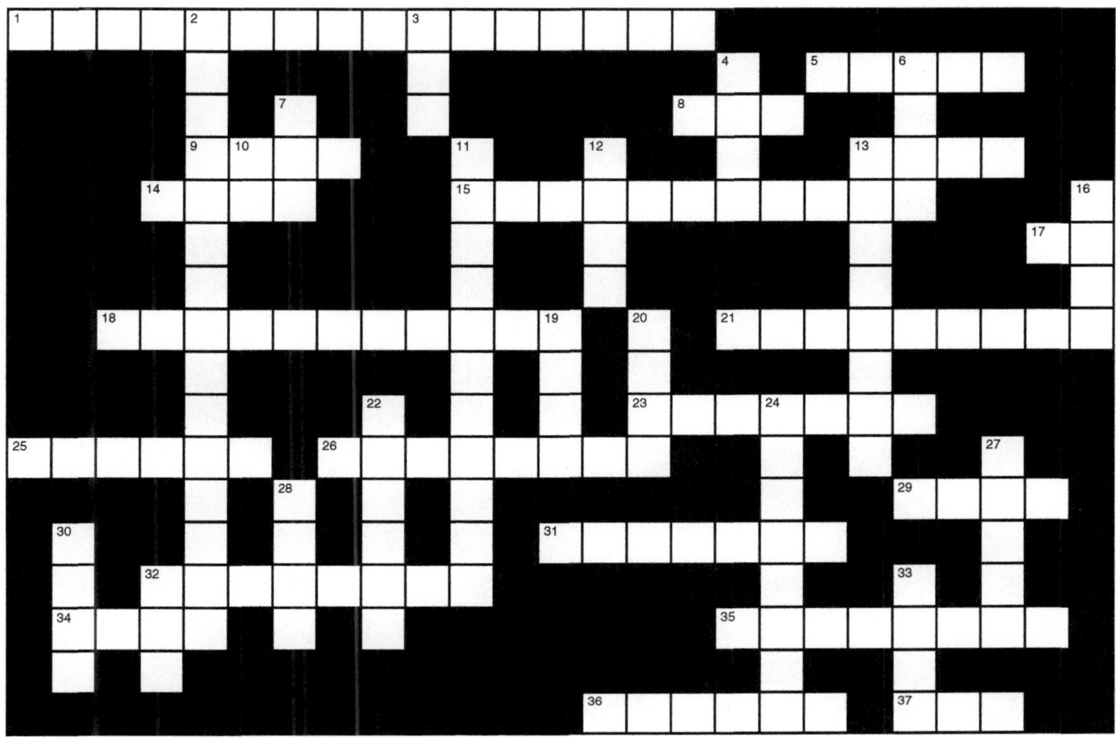

Across

1. Name for local messenger molecule
5. 2nd "b" word in BBB; Blood _____ barrier
8. Charged particle
9. Many of these make up a chromosome
13. A third 21st chromosome results in this syndrome
14. A Greek _____ (e.g., Soma or Glia or Sclerosis)
15. Not a motor or sensory neuron
17. "It's a boy!" . . . . genetically speaking
18. Movement disorder; _____ chorea
21. Appearance of a genotype
23. The "T" from ATCG
25. With neutrons and electrons they make up 28 down
26. Neuron lipid bilayer
29. Neuron or glial, e.g.
31. 3rd "b" word in BBB; Blood _____ _____
32. Found on 10 down, they build proteins
34. Special form of 32 down; abbr.
35. Myelin degeneration; _____ sclerosis
36. Researching monk
37. Genetic blueprint; abbr.

Down

2. Type of CNS glial cell
3. X and Y determine this
4. Unmyelinated segment, with Ranvier
6. May be myelinated or unmyelinated
7. Another name for axon terminal; _____ foot
10. Endoplasmic reticulum; abbr.
11. Organelle used for intracellular transport
12. Creator of the robotic cricket
13. This gene, not recessive, determines 21 across
16. Most common allele in the population; wild-_____
19. "Body" as in a neuron
20. One function of membrane embedded protein; ion _____
22. Cell with axons and dendrites
24. Smallest unit of any substance
27. Neurons and these cells make up your brain
28. See 25 across
30. One function of membrane embedded protein: sodium/potassium _____
32. Ribonucleic acid; abbr.
33. Term for genetic control animal; "_____ type"

CHAPTER

4 How Do Neurons Use Electrical Signals to Transmit Information?

CHAPTER SUMMARY

It has been known for some time that electricity is intimately involved in the process of neural communication. Among the earliest findings supporting this notion was that *electrical stimulation* of a motor neuron resulted in contraction of the muscle that it innervated. Prompted by this finding, researchers began utilizing *electrodes* coupled with electrical stimulators to assess the role of electricity in nerve neural communication. Similarly, some experiments have utilized *recording electrodes* to measure changes in electrical activity associated with neural activity. Perhaps the simplest example of a recording-electrode method is an *EEG*, which is used to measure gross changes in electrical brain waves. Early in the history of these studies, researchers realized that although the brain generates electricity (as seen with an EEG) and that electricity may stimulate neural activity, the speed of neural activity was far too slow to be explained as a purely electrical process. Thus, it was reasoned that electricity is a necessary component of neural communication, but not necessarily the means by which all neurons communicate.

Much of what we know about the electrical component of neural communication has been discovered through research using the *giant axon* of the North Atlantic squid. *Alan Hodgkin* and *Andrew Huxley* were pioneers in this field, utilizing stimulating and recording *microelectrodes* in combination with an *oscilloscope* to study electrical changes across the neural membrane. Such electrical changes are a product of positively and negatively charged *ions* flowing into and out of the cell. It has been established that *diffusion*, *concentration*, and *charge* all influence the movement of ions (and thus the electrical charge of a cell). Diffusion is the process of molecules moving passively and randomly until dynamic equilibrium is reached. Concentration gradient is the orderly movement of molecules from an area of high concentration to an area of low concentration. Charge (or voltage gradient) is the movement of ions toward ions with an opposing charge. These three factors are all mediated by particular features of the neural membrane. For example, *protein channels* may act as pores, allowing passive free movement of molecules. Other proteins may act as *gated channels*, allowing ions to enter and exit the cell at certain times. Still other proteins actively expel some ions, while recruiting others into the cell.

When a neuron is at its *resting potential* there is an electrical imbalance between the intracellular environment and the extracellular environment that is actively maintained.

This *transmembrane voltage* is usually a difference of around *70 mV*, with the intracellular environment more negatively charged. The negative intracellular environment results from *negatively charged proteins* that are too large to pass through any membrane pores. These negatively charged proteins in turn attract a large number of positively charged *potassium ions* into the cell. However, when the concentration gradient inside the cell prevents the entry of additional potassium ions, the cell still maintains a relatively strong negative charge (−70 mV). Positively charged *sodium ions* in high concentration outside of the cell are attracted to the intracellular environment by both electrical and concentration gradients. However, sodium channels are actively gated shut during the resting potential, preventing the entry of these ions. Finally, the cell's *sodium–potassium pump* (another membrane protein) actively works to maintain sodium and potassium imbalances by moving sodium ions from inside the cell to outside, and potassium from outside to in. Negatively charged *chloride ions* contribute very little to this process. The higher extracellular concentration drives these ions toward the intracellular environment, but this force is balanced by the negative charge inside the cell repelling the influx.

Graded potentials are seen when the cell potential becomes more positive (*depolarization*) or more negative (*hyperpolarization*) in a graded fashion through small voltage fluctuations. Such a change may be evoked with a small current from a stimulating electrode. Graded potentials occur only in the membrane vicinity near the stimulation, and they decay in magnitude as they move from the point of stimulation. Depolarization generally results from sodium ions flowing into a neuron, while hyperpolarization results from potassium ions flowing out of a neuron. *Action potentials* (nerve impulses) are evoked when a neuron is depolarized to a *threshold potential*. Threshold is around −50 mV (compared to resting potential of −70 mV). At threshold the graded nature of the depolarization changes dramatically. The cell membrane rapidly moves in a positive direction, past *equilibrium* (0 mV) to a positive charge of approximately +30 mV, and then quickly returns to its *polarized* state of −70 mV. This brief fluctuation can be explained physiologically by an opening of *voltage-sensitive* sodium and potassium channels at threshold. Sodium gates, which open first, allow a large influx of positively charged ions moving with the force of both concentration and electrical gradients into the cell. Potassium channels open just slightly slower (as the action potential approaches its peak of +30 mV), allowing potassium to flow out of the cell with the force of these same two gradients. A final interesting feature of the action potential is the concept of the *refractory period*. The *absolute refractory* period occurs during an action potential, when sodium and potassium gates are open. During this time it is impossible to stimulate a second action potential. The *relative refractory* period occurs during a brief time after the action potential when the cell is slightly hyperpolarized. During this period, a second action potential may be evoked, but the membrane charge required to reach threshold is greater than that needed during the resting potential.

A nerve impulse is simply the movement (or propagation) of an action potential along the axon of a neuron. The initiation of an action potential anywhere on a cell necessarily changes the membrane potential around that region. With sufficient depolarization, surrounding voltage-sensitive channels will open, initiating a propagation of the signal. The refractory period (mentioned above) prevents an action potential from moving backward over a region that has already been depolarized. This results in propagation in a single direction, usually thought of as along the axon from the cell body to the terminal. With this in mind, an action potential (a nerve impulse) initiated at the juncture of the cell body and the axon (called the axon hillock) will move slowly along the length of the axon until reaching the terminal. This process occurs in *unmyelinated axons*, but is a relatively slow method of propagation. The vast majority of our cells incorporate the use of a *myelin sheath* that acts as an insulator around axons, speeding propagation. Specifically, propagation is speeded by large numbers of sodium channels clustering in unmyelinated portions of the axon (called *nodes of Ranvier*) and forcing the electrically charged particles to "jump" under myelinated sections of the axon that lack channels. This jumping is called *saltatory conduction*, a process that vastly speeds the neural communication process.

When a neuron receives synaptic input from innervating terminals, this input may be excitatory (*EPSP*), causing membrane depolarization, or inhibitory (*IPSP*) causing membrane hyperpolarization. Most cells receive many inputs that are combinations of EPSPs and IPSPs. In this regard the neuron must weigh, or *integrate*, these inputs. Integration can be thought of in terms of summation of signals. Signals may be summated from numerous sites in close spatial proximity (*spatial summation*) or they may be summated in time from numerous inputs from the same site in close temporal proximity (*temporal summation*). If summation of excitatory input is sufficient to bring the cell membrane to threshold, an action potential is stimulated. Action potentials are most frequently initiated at the axon hillock where voltage-dependent sodium channels are numerous and sometimes more sensitive to membrane changes than other regions on the neural membrane.

The process of producing neural signals from environmental stimuli (sight, sound, touch, etc.) requires specialized neural receptors capable of transducing environmental energy into neural energy. Likewise, the process of producing a behavior from a neural signal requires stimulation of muscle fibers with neurotransmitters. One example of this is the movement of striated muscles, which is produced by releasing the neurotransmitter *acetylcholine* onto the muscle *end plate*, which in turn contains *transmitter-activated channels*. When these channels are opened in sufficient number, they produce muscle contraction necessary for movement.

KEY TERMS

The following is a list of important terms introduced in Chapter 4. Give the definition of each term in the space provided.

Electricity and the Neurons

Epilepsy

Negative pole

Positive pole

Volts

Electrical potential

Potential

Current

Stimulating electrode

Voltmeter

Recording electrode

Electroencephalogram, or EEG

Tools for Measuring Elecricity in Neurons

Giant axon

Oscilloscope

Milliseconds

Millivolts

Microelectrodes

Movement of Ions

Diffusion

Concentration gradient

Voltage gradient

Electrical Activity of a Membrane

Transmembrane voltage

Resting potential

Sodium–potassium pump

Graded Potentials

Hyperpolarization

Depolarization

Tetraethylammonium (TEA)

Tetrodotoxin

Action Potential

Threshold potential

Voltage-sensitive channels

Absolute refractory

Relative refractory

Nerve Impulse

Glial cells

Schwann cells

Oligodendroglia

Nodes of Ranvier

Saltatory conduction

Myasthenia gravis

Integrating Information

Excitatory postsynaptic potential, or EPSP

Inhibitory postsynaptic potential, or IPSP

Axon hillock

Temporal summation

Spatial summation

Stretch-sensitive channels

Transducing

End plate

Acetylcholine

Transmitter-sensitive channels

KEY NAMES

The following is a list of important names introduced in Chapter 4. Explain the importance of each person in the space provided.

Luigi Galvani

Roberts Bartholow

Hermann von Helmholtz

Alan Hodgkin and Andrew Huxley

PRACTICE TEST

Multiple-Choice Questions

Answer each of the following multiple-choice questions with the best possible answer based on information from your text.

1. Early studies of the effects of electricity on brain function were sometimes conducted in awake humans with exposed brain tissue. For example, R. Bartholow stimulated the cortex of a patient and reported which of the following?
   A. He was able to control her speech by stimulating specific parts of the brain.
   B. He was able to evoke memories by stimulating specific parts of the brain.
   C. He was able to evoke hand movements by stimulating specific parts of the brain.
   D. He was able to evoke fear and anxiety by stimulating specific parts of the brain.
   E. All of the answers are correct.

2. When recording electrical potentials from the brain, one wire from a voltmeter is attached to a recording electrode. Which of the following is also connected to the voltmeter?
   A. a battery
   B. a stimulating electrode
   C. a reference electrode
   D. a brain cell
   E. the skull

3. Although it was initially believed that electricity progressed as a continuous electrical signal along the nerve pathways, it was later determined that the speed of neural transmission is slower than the speed of electricity. In 1886 Julius Bernstein suggested the slowing of electrical transmission was caused by which of the following?
   A. myelin
   B. a chemical basis for the electrical charge
   C. iron in the blood
   D. density of nerve cells
   E. gravitational forces

4. The giant axon of the squid has become a popular tool for studying neural transmission and electrical potentials for which of the following reasons?
   A. It is very large, and therefore easily manipulated for study.
   B. It is a very slow firing axon, making it easy to study.
   C. It is approximately 12 feet long, making it useful for studying transmission patterns.
   D. It can be kept functional in a liquid bath of any fluid.
   E. All of the answers are correct.

5. An oscilloscope is:
   A. similar in design to a television.
   B. capable of being used as a sensitive voltmeter.
   C. a useful tool for studying electrical potential in neurons.
   D. capable of indicating direction, duration, and magnitude of electrical changes.
   E. All of the answers are correct.

6. What is the reason that negatively charged proteins inside the cell do *not* move to the extracellular fluid when the cell is at rest?
   A. They are kept in the cell by the concentration gradient.
   B. They are kept in the cell by the voltage gradient.
   C. They are kept in the cell by both the voltage and the concentration gradient.
   D. Channels for these proteins only open during the initiation of an action potential.
   E. None of the answers is correct.

7. Movement down a concentration gradient describes which of the following?
   A. movement of ions from an area of high concentration to an area of low concentration
   B. movement of ions from an area of low concentration to an area of high concentration
   C. movement of positively charged ions toward an area with a net negative charge
   D. movement of positively charged ions toward an area with a net positive charge
   E. All of the answers are correct.

8. Which of the following best describes the resting potential of a typical neural membrane?
   A. The inside charge is approximately equal to the outside charge.
   B. The inside charge is approximately 70 volts less than the outside charge.
   C. The inside charge is approximately 70 volts greater than the outside charge.
   D. The inside charge is approximately 70 millivolts less than the outside charge.
   E. The inside charge is approximately 70 millivolts greater than the outside charge.

9. The number of potassium ions that can accumulate in a resting neuron is restricted by the concentration gradient. What is the primary force preventing sodium ions from accumulating in a resting neuron?
   A. concentration gradient
   B. voltage gradient
   C. both concentration and voltage gradients
   D. the sodium–potassium pump
   E. entry restricted by gated channels at rest

10. Which of the following most accurately describes the sodium–potassium pump?
   A. It is a protein molecule imbedded in the cell membrane.
   B. It is inactive during the resting state of the neuron.
   C. It moves sodium ions into the cell during rest.
   D. It moves potassium ions out of the cell at rest.
   E. All of the answers are correct.

11. Which of the following best describes graded potentials?
   A. They are not all-or-none phenomena.
   B. They appear as a movement of the membrane potential from positive to negative.
   C. They appear as a movement of the membrane potential from negative to positive.
   D. They may be produced by the opening of sodium channels.
   E. All of the answers are correct.

12. Which of the following phenomena occur when a membrane reaches threshold potential?
   A. voltage-sensitive potassium channels close
   B. voltage-sensitive sodium channels close
   C. voltage-sensitive sodium channels open
   D. voltage-sensitive chloride channels open
   E. voltage-sensitive chloride channels close

13. While an action potential is occurring a neuron is said to be in an absolute refractory period, during which time it is incapable of initiating another action potential. For a brief period immediately following the action potential the neuron enters a relative refractory period. Which of the following statements best describes the relative refractory period?
   A. The cell is slightly depolarized compared with normal resting potential.
   B. Initiation of an action potential requires slightly greater stimulus intensity at this time.
   C. Initiation of an action potential requires slightly less stimulus intensity at this time.
   D. The cell is incapable of initiating another action potential at this time.
   E. None of the answers is correct.

14. The term *saltatory conduction* is derived in part from Latin to describe which of the following?
    A. the movement of a signal in a single direction
    B. the rapid conduction of a signal along an axon
    C. the jumping action of a signal along an axon
    D. the opening of voltage-sensitive channels required for an action potential
    E. the closing of voltage-sensitive channels following an action potential

15. Which of the following is *not* a feature of nodes of Ranvier?
    A. They are an unmyelinated portion of the dendrite.
    B. They contain a high density of sodium channels.
    C. They are necessary for signal propagation via saltatory conduction.
    D. They are necessary for rapid propagation of a neural signal.
    E. All of the answers are correct.

16. Which of the following would be most likely to evoke an IPSP?
    A. an influx of sodium ions
    B. an influx of potassium ions
    C. an influx of both potassium and sodium ions
    D. an influx of chloride ions
    E. an efflux of chloride ions

17. If an IPSP were to occur in close proximity on a neural membrane to an EPSP, what would be the net result?
    A. They would summate to produce a greater depolarization than either one separately.
    B. They would summate to produce a greater hyperpolarization than either one separately.
    C. They would produce approximately the same depolarization that either would individually.
    D. They would produce approximately the same hyperpolarization that either would individually.
    E. They would act as opposing forces, each canceling the membrane effect of the other.

18. Which of the following describes the axon hillock?
    A. It is found at the junction where the axon meets the terminal button.
    B. It is generally affected more by EPSPs and IPSPs initiated at the dendrites than on the cell body.
    C. It contains a high concentration of voltage-sensitive chloride channels.
    D. It is where most action potentials are initiated.
    E. All of the answers are correct.

19. Some dendrites specialized for sensory detection are wrapped around the base of hair cells. These dendrites conduct a nerve impulse when:
    A. stimulated by neurochemicals.
    B. warmed.
    C. vibrated.
    D. stretched.
    E. exposed to high concentrations of sodium.

20. Which of the following neurotransmitters stimulates transmitter-activated channels at muscle fibers?
    A. dopamine
    B. serotonin
    C. acetylcholine
    D. norepinephrine
    E. epinephrine

Short-Answer Questions

Answer each of the following questions with a brief but complete written answer based on information from your text.

1. Among the early theories of neural communication was the idea that neurons propagated and transmitted electrical signals. Hermann von Helmholtz refuted this theory with a relatively simple experiment. Briefly describe the techniques he used and the results he generated in this experiment.

2. The oscilloscope has become one of the most useful tools for assessing neural activity. This relatively simple device generates a "line" from which neural activity can be interpreted. In general terms, what does a vertical deflection of this line indicate? What is the difference between a small deflection and a large deflection of this line? What is indicated when the vertical deflection is then maintained as a horizontal line?

3. Briefly explain why negatively charged intracellular proteins do not diffuse out of the cell. Also explain why sodium and potassium ions are attracted to these intracellular proteins.

4. Briefly explain why chloride ions show so little movement into and out of a neuron when it is at its resting potential.

5. Briefly explain why a neuron maintains a negative intracellular charge, even though there is approximately 20 times as many positively charged potassium ions inside the cell relative to outside the cell at this time.

6. Novocaine's function is similar to that of tetrodotoxin (the puffer fish toxin). Briefly describe how tetrodotoxin has its effects, and then speculate on how injection of novocaine may work as a local anesthetic.

7. Briefly explain the difference between the absolute refractory period and the relative refractory period.

8. When an action potential is initiated by membrane depolarization it is propagated by opening adjacent voltage-sensitive membrane channels. Why then does a signal move in only one direction along a membrane, and not reverse direction by opening voltage-sensitive channels from which it came?

9. Explain saltatory conduction, including nodes of Ranvier and the myelin sheath in your description.

10. Compare EPSPs to IPSPs in terms of how they differ in associated influx and efflux of ions, how they affect cell membrane potential, and how they affect cell firing rate.

Matching Questions

Complete each of the following matching questions based on information from your text.

1. Match the following ions to their respective charge.

	___ Potassium
Positive (+)	___ Sodium
Negative (–)	___ Calcium
	___ Chloride
	___ Intracellular proteins

2. Identify each of the following as being either intracellular (I) or extracellular (E) when a neuron is in a resting state.

__ Large negatively charged protein molecules
__ Higher concentration of potassium ions
__ Higher concentration of sodium ions
__ Negative charge
__ Higher concentration of chloride ions

3. Identify each of the following events as occurring with an influx of sodium (IS) or with an efflux of potassium (EP).

__ Initiation of the action potential
__ Hyperpolarization
__ Depolarization
__ Refractory period

4. Match the following disorder with the appropriate symptom or characteristic.

A. Lou Gehrig's disease
B. Myasthenia gravis
C. Epilepsy

___ Insensitivity to the chemical messages
___ Autoimmune disorder
___ Produces abnormal EEG
___ Causes degeneration of motor neurons
___ Neurons fire synchronously

5. Label each of the following features as being exclusive to action potentials (AP), exclusive to graded potentials (GP), or as a feature shared by both types of potentials (AP + GP).

___ Generally changes membrane potential by greater than 50mV
___ May move in a positive or a negative direction
___ Is initiated at a threshold charge
___ May occur when sodium channels open
___ Concludes with a refractory period

Diagrams

1. On the graph below draw an approximate oscilloscope recording from a neuron. Begin by drawing a resting state. At point A illustrate a subthreshold depolarization. At point B illustrate a hyperpolarization. At point C illustrate a threshold depolarization and action potential.

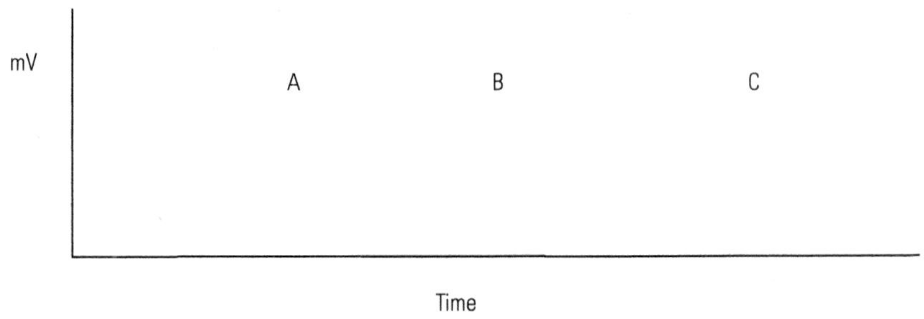

mV

A B C

Time

2. On the graph below draw an approximate oscilloscope recording from a neuron. Begin by drawing a resting state. At point A illustrate the response that would be seen with an efflux of potassium ions. At point B illustrate the response that would be seen with an efflux of sodium ions. At point C illustrate the response that would be seen with a subthreshold influx of calcium ions.

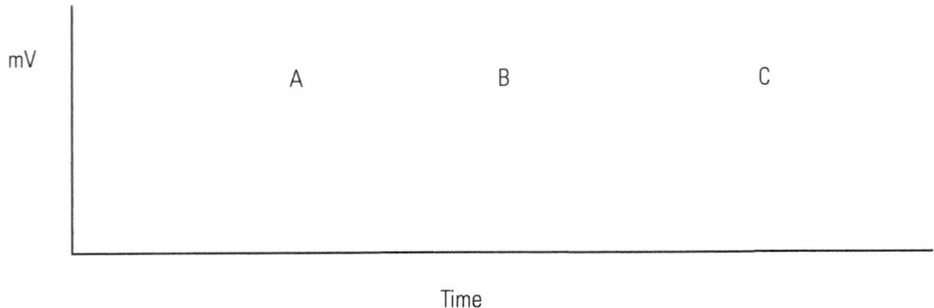

3. Illustrate a section of a membrane that contains each of the following proteins:
   A. potassium channel (in an open state)
   B. sodium channel (in a closed state)
   C. sodium–potassium pump

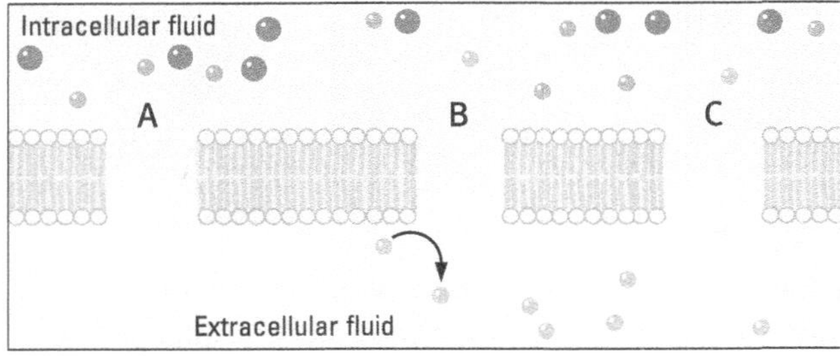

4. A diagram of an axon follows. Add to this diagram a typical pattern of myelin. Indicate nodes of Ranvier. Also indicate where you would expect sodium channels to be concentrated.

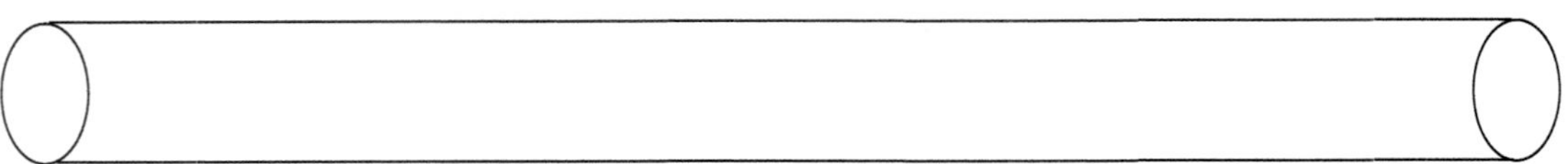

5. Assuming all other variables are held constant, which of the two neurons is more likely to be activated based on the inputs indicated?

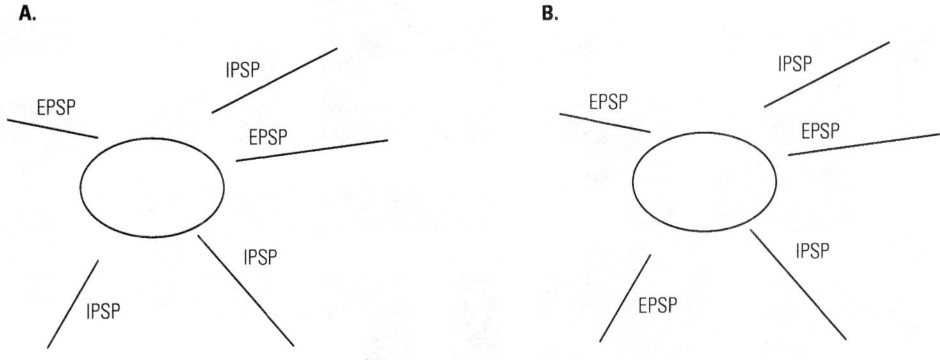

The Web

Consider using the following Web sites for additional information on some of the topics from this chapter:

1. Neural activity animated and interactive: icarus.med.utoronto.ca/neurons/index.swf
2. Epilepsy Foundation of America: www.epilepsyfoundation.org/
3. World Federation of Neurology Amyotrophic Lateral Sclerosis: www.wfnals.org/
4. Myasthenia Gravis Foundation of America: www.myasthenia.org/
5. How to use an oscilloscope: www.doctronics.co.uk/scope.htm

CROSSWORD PUZZLE

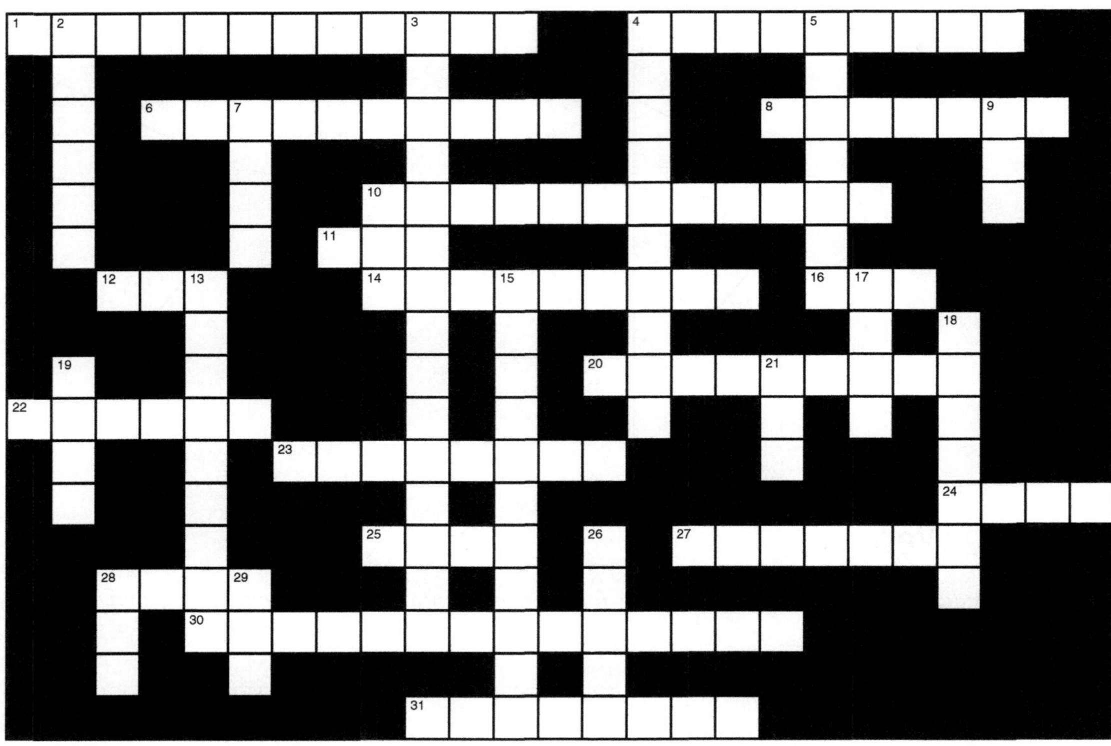

Across

1. Used as a sensitive voltmeter to measure neural activity
4. It flows out at the action potential peak
6. Movement of ions creates this type of activity
8. Proper name for unmyelinated region is node of _____
10. Term for neuron receiving a signal from another neuron
11. Acetylcholine released here on muscle surface; _____ plate
12. The Na+–K+ protein pump moves Na+ _____
14. Most channels are voltage-_____, some are stretch-_____
16. Phenomenon in 10 down is actually termed a _____
20. Device used for 15 down
22. Pioneers in squid axon research . . . Hodgkin and _____
23. Refractory period that is not relative
24. Inhibitory signal; abbr.
25. 27 across and 3 down; e.g.
27. Peripheral myelin cells
28. Unit of measure for cell activity, with milli
30. 2 down is an example of this
31. Hyperpolarization is this direction

Down

2. ALS affects primarily _____ motor neurons
3. CNS cells that insulate axons
4. Graded and action, e.g.
5. Signal summation may be temporal or _____
7. Excitatory signal below 13 down
9. Used to record brain activity from scalp
10. System where Schwann cells are found; abbr.
13. Point where action potential is initiated
15. Neurons are studied by recording and _____
17. What 27 cross and 3 down insulate
18. Potential between 24 across and 7 down
19. Protein that moves sodium and potassium
21. # of ions actively pumped at rest
26. Describes squid axons used for research
28. Sodium is pumped out ___ active transport
29. A refreshing drink . . . or tetraethylammonium, for short

CHAPTER

5 How Do Neurons Use Electrochemical Signals to Communicate and Adapt?

CHAPTER SUMMARY

It has been known for nearly a century that neurons communicate with target structures by releasing chemical *neurotransmitters* at their synapses. Beginning in the 1950s, use of the electron microscope provided images of the structural basis for chemical communication. These images show that chemicals are packaged in *synaptic vesicles*, and that vesicles may be contained within *storage granules* in the presynaptic terminal. Electron microscope images also show a variety of protein molecules in both the *presynaptic* and *postsynaptic membranes*. These proteins act as receptors, channels, and pumps, all utilized in neural communication. Although the vast majority of mammalian synaptic connections use chemicals for signal transmission, in some rare cases *electrical synapses* (commonly referred to as *gap junctions*), can be found. Chemical communication allows greater flexibility than electrical signals in transmitting a signal at the synapse; however, this flexibility comes at the cost of speed. Chemical transmission is slower than electrical in part because of the functional stages required, which include 1) synthesis and storage of chemical molecules; 2) transportation from the cell body to the terminal and from the terminal to the presynaptic membrane at the time of release; 3) interaction of chemicals with postsynaptic membrane receptor sites; and 4) deactivation of transmitter after use. Some neurotransmitters are synthesized from food, while others are simply made from instructions inherent in DNA coding. From the storage vesicles, neurotransmitter is released when an influx of calcium at the terminal binds with *calmodulin* and the resulting calcium/calmodulin complex moves vesicles to the presynaptic membrane. The strength of a neural signal depends on how much calcium enters a terminal and subsequently how many vesicles of neurotransmitter are released into the *synaptic cleft*. The contents of a single vesicle are called a *quantum*. A quantum constitutes the smallest amount of transmitter that can be released, and all neural signals constitute some multiple of this quantum.

Once released into the synaptic cleft, neurotransmitters bind to *transmitter-activated receptors* in the postsynaptic membrane, where they may hyperpolarize or depolarize that cell. In some cases, binding to receptor sites may initiate a chain of chemical reactions in the postsynaptic cell. Deactivation of transmitter once released may result from passive diffusion away from receptors, degradation by enzymes in the cleft, reuptake into the presynaptic terminal, or uptake by neighboring glial cells. Enzymatic degradation may also take place inside the terminal when a neuron produces more neurotransmitter than is needed.

Synaptic connections are generally thought of as occurring between axon terminals and dendrites. Although such *axodendritic* synapses are probably the most common, there are also *axoaxonic*, *axosynaptic*, *axomuscular* (terminating on muscle fibers), *axoextracellular* (terminating in extracellular space), and *axosecretory* (terminating on blood capillaries). In some cases dendrites synapse on other dendrites (*dendrodendritic*). Regardless of location, all synapses are categorized as either *Type I* (excitatory) or *Type II* (inhibitory) based on the effect they ultimately have on the postsynaptic neuron. Although Type I synapses may intuitively seem "important" in mediating behavior, it should be realized that Type II synapses are equally important in regulating normal function. For example, in neural disorders where Type II synapses are dysfunctional, loss of inhibitory input may result in uncontrolled movements such as *tremors* or *dyskinesia*.

The earliest studies to suggest chemical communication within the nervous system assumed the existence of an excitatory and an inhibitory neurotransmitter combining to mediate behavior. This assumption was correct but oversimplified. Currently, about 50 different substances have been identified that act as neurotransmitters. It is also likely that many such substances are still unidentified. In general terms, for a substance to be considered a neurotransmitter, it should meet four criteria: 1) It must be synthesized in, or present in, a neuron; 2) When the neuron is active it releases the substance which in turn evokes a response; 3) When the substance is experimentally administered at the target, the same response is evoked; and 4) There must be a mechanism for deactivation of the substance in the synaptic cleft. However, even these criteria are considered too strict by some researchers, since a very wide range of substances appear to have some role in mediating neural function. The term *putative neurotransmitter* is used to describe substances suspected, but not yet confirmed, to be true neurotransmitters.

Currently, neurotransmitters are classified as *small-molecule neurotransmitters*, *peptide transmitters*, and *gases*. Small-molecule neurotransmitters (or their main components) are derived from the food we eat. They are synthesized and packaged in the terminal. They act relatively quickly at target sites and are replaced quickly after use. Many psychoactive drugs are designed to reach the brain in a manner similar to small-molecule neurotransmitters. Examples of small-molecule neurotransmitters include *acetylcholine* (*ACh*, made from *choline* and *acetate*), *dopamine* (made from tyrosine), *norepinephrine* (made from dopamine), and *epinephrine* (made from norepinephrine). *Gamma-aminobutyric acid* (*GABA*) and *glutamate* are the primary inhibitory and excitatory brain neurotransmitters, respectively (although *glycine* is also a major inhibitory neurotransmitter at many structures). Unlike small-molecule neurotransmitters, most peptide neurotransmitters (or *neuropeptides*), are assembled and packaged in the soma and then transported to the axon terminal. This process makes replacement after use relatively slow. In addition, peptide transmitters do not act directly at ion channels, but rather, produce effects through indirect influence of cell structure and function. *Enkephalins* are one example of peptide transmitters. These substances are known to regulate pain, an effect that can be seen in the response to opium and morphine, both of which mimic the actions of *Met-enkephalin*, *Leu-enkephalin*, and *beta-endorphin*. *Nitric oxide* (*NO*) and *carbon monoxide* (*CO*) are gases that act as neurotransmitters. Neither stored in nor released from vesicles, these substances appear to be synthesized in many regions of the cell and then diffuse freely across the membrane as needed.

There are two general classes of receptor sites to which neurotransmitters bind. *Ionotropic* receptors are functionally linked to ion channels, and when activated produce rapid changes in ionic flow and membrane potential of the target cell. *Metabotropic* receptors are associated with one of a family of proteins called *guanyl nucleotide-binding proteins* (*G-proteins*). When these sites are activated they trigger a series of changes within the target cell utilizing a second messenger system. The *second messenger* in turn may alter ion flow through membrane channels. It may also result in production of new channels or the production of other new proteins via a message to the cell's nuclear DNA. Target cells may contain exclusively ionotropic or metabotropic receptor sites, or a combination of the two. Furthermore, these receptors may exert any combination of excitatory or inhibitory influence on the target cell. It is this potential combination of inputs that generates the flexibility of an infinite number of possible responses to individual stimuli.

It was originally believed any given neuron contained only one neurotransmitter substance. This hypothesis has since been disproved with the finding that many combinations of transmitters may coexist within a given neuron or even within a given axon terminal. This coexistence of transmitter substances, in combination with the large number of identified transmitters, and the potential for excitatory or inhibitory postsynaptic effects results in a staggering number of possible combinations for activity within the nervous system.

The skeletal motor system utilizes *cholinergic* neurons releasing ACh at the neuromuscular junction. When stimulated, *nicotinic ACh receptors* (*nACh*) permit simultaneous efflux of potassium ions and influx of sodium ions, producing an excitatory response. The *autonomic nervous system* contains both *cholinergic* and *adrenergic* neurons. Adrenergic neurons release epinephrine (also called adrenaline) to stimulate the body into a "fight or flight" response. The release of acetylcholine produces the opposite response, sometimes termed the "rest and digest" response. The *central nervous system* utilizes a wide array of transmitter substances, each with a specific function. As mentioned, GABA and glutamate have inhibitory and excitatory effects, respectively, throughout this system. In addition, four other neurotransmitter systems exert major influence in what are referred to as the *activating systems*. These systems are the cholinergic, dopaminergic, noradrenergic, and serotonergic systems. In very general terms, the cholinergic system is associated with normal alert behaviors including learning and memory. Degeneration of this system is a feature of Alzheimer's disease. The dopaminergic system controls movement and has been implicated in drug addiction and schizophrenia. The noradrenergic system is associated with mania when overactive, and depression when underactive. The serotonergic system is also associated with depression, some forms of schizophrenia, and obsessive-compulsive disorder.

It is now known that changes in synaptic function, including release and response to neurotransmitters, are associated with learning. Pioneering studies by *Kandel* and colleagues using *Aplysia* have shown, for example, that *habituation* results from a reduction in neurotransmitter release associated with repeated stimulation. This reduced transmitter release appears to result from reduced calcium influx at the terminal. However, it is still unknown why calcium channels exhibit this change in the habituation process. *Sensitization*, the learned behavioral response opposite that of habituation, appears to result from serotonin released by interneurons in *Aplysia*. Acting as a second messenger, serotonin increases the response of the target neuron, evoking an exaggerated postsynaptic response. Researchers have found that some cells, particularly those in the hippocampus, exhibit functional changes in response to stimulation, which may be apparent for several weeks. Such *long-term potentiation* (*LTP*) can occur when glutamate receptors are stimulated as the cell membrane is depolarized. This state of high excitation results in calcium influx that is thought to act as a second messenger producing conformational changes in both the presynaptic and the postsynaptic membrane. One of these changes is production of retrograde plasticity factor that is thought to ultimately stimulate further release of glutamate. This self-perpetuating process eventually leads to relatively permanent changes. *Associative* learning is thought to occur when two stimuli are paired in time to create a neurotransmitter response vigorous enough to evoke LTP synaptic changes. Finally, recent research has shown dendritic processes to be very plastic. That is, they have the potential to increase or decrease surface area and spiny projections. The ability to undergo such physical changes further increases the potential for synaptic conformations to underlie the process of learning.

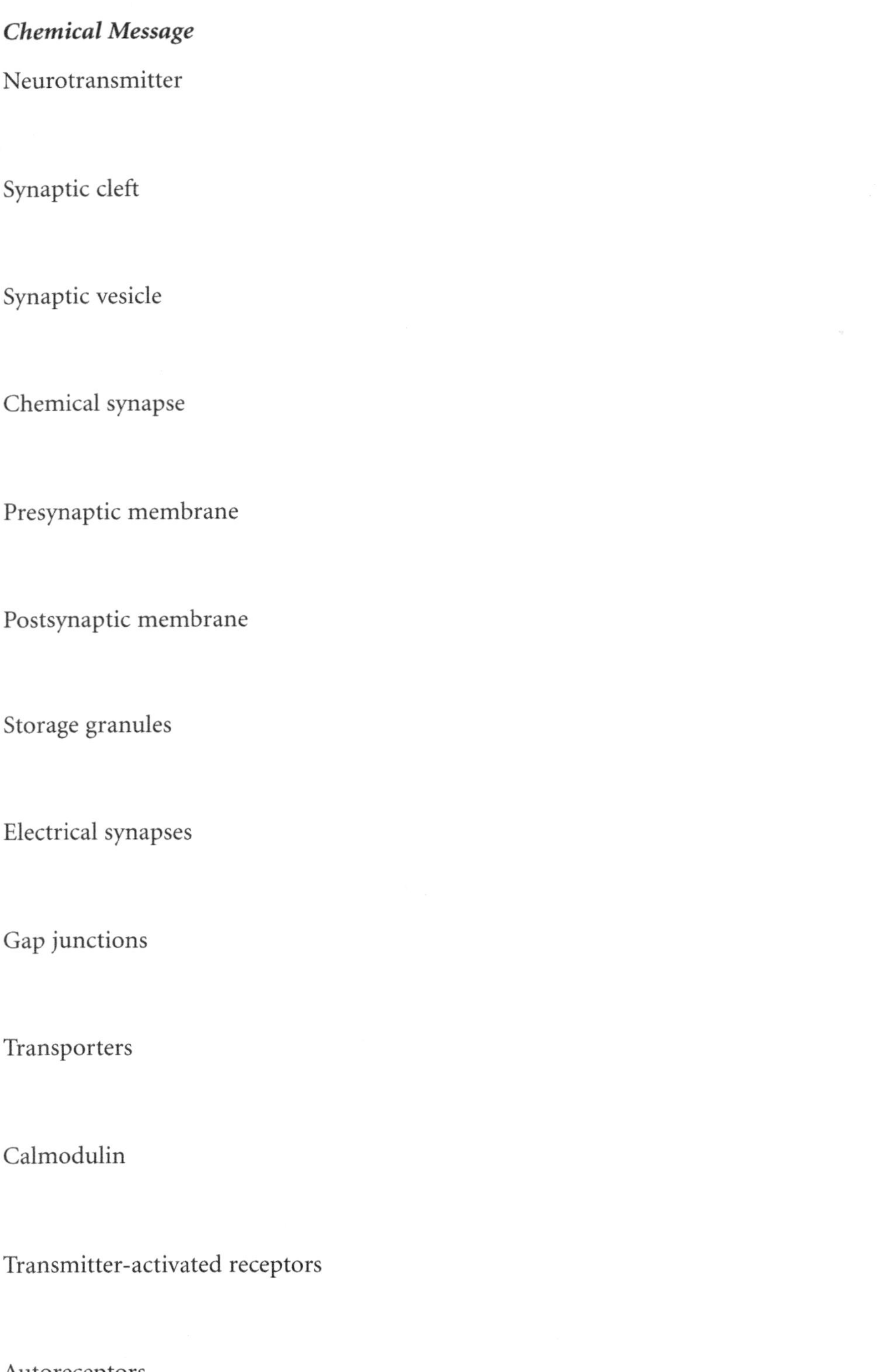

KEY TERMS

The following is a list of important terms introduced in Chapter 5. Give the definition of each term in the space provided.

Chemical Message

Neurotransmitter

Synaptic cleft

Synaptic vesicle

Chemical synapse

Presynaptic membrane

Postsynaptic membrane

Storage granules

Electrical synapses

Gap junctions

Transporters

Calmodulin

Transmitter-activated receptors

Autoreceptors

Quantum

Reuptake

Axodendritic

Axomuscular

Axosomatic

Axoaxonic

Axosynaptic

Dendrodendritic

Axoextracellular

Axosecretory

Type I synapses

Type II synapses

Varieties of Neurotransmitters

Small-molecule transmitters

Putative neurotransmitter

Acetylcholine (ACh)

Histamine

Choline

Acetate

Dopamine (DA)

Norepinephrine (NE)

Epinephrine (EP)

Rate-limiting factor

Glutamate (Glu)

Serotonin (5-HT)

Gamma-aminobutyric acid (GABA)

Neuropeptide

Met-enkephalin

Leu-enkephalin

Beta-endorphin

Nitric oxide (NO)

Carbon monoxide (CO)

Two Classes of Receptors

Ionotropic

Metabotropic

Guanyl nucleotide-binding (G-proteins)

Second messenger

Neurotransmitter Systems

Cholinergic

Nicotinic ACh receptor (nACh)

Fight-or-flight response

Rest-and-digest response

Activating systems

Cholinergic system

Dopaminergic system

Noradrenergic system

Serotonergic system

Alzheimer's disease

Nigrostriatal dopamine system

Mesolimbic dopamine system

Major depression

Schizophrenia

Obsessive-compulsive disorder (OCD)

Synapses and Learning

Neuroplasticity

Learning

Habituation

Associative learning

Long-term potentiation (LTP)

NMDA

AMPA

Long-term depression (LTD)

KEY NAMES

The following is a list of important names introduced in Chapter 5. Explain the importance of each person in the space provided.

Otto Loewi

Bernard Katz

Donald Hebb

Eric Kandel

PRACTICE TEST

Multiple-Choice Questions

Answer each of the following multiple-choice questions with the best possible answer based on information from your text.

1. Neurotransmitter mediation of heart rate is one example of chemical communication regulating behavior. When the puffin dives beneath the surface of the water it undergoes a behavior called diving bradycardia. What is the functional purpose of this neurotransmitter-mediated response?
   A. to increase oxygen consumption when not breathing
   B. to conserve oxygen while underwater
   C. to increase heart rate
   D. to increase metabolism and raise body temperature
   E. The reason for this response is unknown.

2. Early researchers, including Loewi, determined that ___________ increased heart rate, whereas ____________ decreased heart rate.
   A. acetylcholine; epinephrine
   B. epinephrine; acetylcholine
   C. electricity; acetylcholine
   D. electricity; epinephrine
   E. electricity; lack of electricity

3. Neurotransmitter is known to be stored in synaptic vesicles. However, synaptic vesicles may also be stored in larger vesicles. What is the name given to these large storage vesicles?
   A. Golgi apparatuses
   B. terminal vesicles
   C. giant vesicles
   D. storage granules
   E. terminal storage units

4. Which of the following ions plays an important role in transporting vesicles to the membrane and in the release of neurotransmitters into the synaptic cleft?
   A. sodium
   B. potassium
   C. chloride
   D. magnesium
   E. calcium

5. The process of presynaptic neurotransmitter reuptake requires which of the following?
   A. autoreceptors
   B. storage granules
   C. calcium-calmodulin complex
   D. protein transporter
   E. All of the answers are correct.

6. It could be said that glutamate is generally found at:
   A. Type I synapses.
   B. Type II synapses.
   C. gamma-aminobutyric acid synapses.
   D. axoaxonic synapses.
   E. all synapses in approximately equal frequency and amount.

7. Type I synapses are typically located on the ______________, whereas Type II synapses are usually found on the __________________.
   A. dendrites; axon terminals
   B. dendrites; cell body
   C. axon terminals; dendrites
   D. cell body; dendrites
   E. cell body; axon terminals

8. Acetylcholine (ACh) can always be found at which of the following types of synapses?
   A. axodendritic
   B. axosomatic
   C. axosecretary
   D. axomuscular
   E. dendrodendritic

9. The Renshaw loop describes which of the following?
   A. a device designed to loop around axons and isolate terminals for experimental studies
   B. the process of synthesizing and then breaking down neurotransmitter substances
   C. a feedback loop by which a neuron inhibits itself from continued transmitter release
   D. the sequence of motor neuron, interneuron, sensory neuron, back to interneuron
   E. the process of excitation followed by inhibition found at most synapses

10. Neurotransmitters that are synthesized and packaged for use in the axon terminal also have what feature?
    A. they are generally synthesized from substances in the food we eat
    B. they are usually proteins
    C. they are relatively slow to act at target sites
    D. they are relatively slow to be replaced after use
    E. None of the answers is correct.

11. Regarding the neurotransmitter acetylcholine, which of the following is *not* considered a substance used in the synthesis of this chemical?
    A. choline
    B. acetate
    C. acetylcholinesterase (AChE)
    D. a compound that is found in vinegar
    E. a compound that is found in egg yolk

12. Glycine is considered the primary inhibitory neurotransmitter in the brain stem and spinal cord. Which of the following is considered the primary inhibitory neurotransmitter in regions such as the forebrain and cerebellum?
    A. glutamate
    B. gamma-aminobutyric acid (GABA)
    C. acetylcholine (ACh)
    D. epinephrine (EP)
    E. glycine

13. Opium and morphine are drugs that mimic the effects of several peptide transmitters. Which of the following is *not* mimicked by these drugs?
    A. met-enkephalin
    B. leu-enkephalin
    C. beta-endorphin
    D. alpha-endorphin
    E. These drugs mimic the effects of all peptide transmitters.

14. Which of the following is *not* true of ionotropic receptor sites?
    A. They mediate rapid change.
    B. They are functionally linked to a membrane channel.
    C. They are structurally similar to voltage-sensitive channels.
    D. When activated they allow ions to flow across the membrane.
    E. When activated they generally alter cell metabolism through a series of steps.

15. The nicotinic ACh receptor is located at all neuromuscular junctions on skeletal muscles. Which of the following is true of this receptor?
    A. Stimulation permits an influx of sodium ions.
    B. Stimulation permits an efflux of potassium ions.
    C. This receptor can be stimulated by nicotine.
    D. Effects of ACh at the receptors can be altered by calcitonin gene-related peptide (CGRP).
    E. All of these statements are true of this receptor.

16. As an example of how loss of neurotransmitter function may affect behavior, it is known that humans with Alzheimer's disease show extensive loss of ____________ neurons.
    A. dopaminergic
    B. serotonergic
    C. cholinergic
    D. opiate
    E. All of the answers are correct.

17. Which of the following is true of the neural basis of learning the habituation response as shown in experiments with *Aplysia*?
    A. Ability to produce action potentials is reduced in sensory neurons.
    B. Ability to produce action potentials is reduced in motor neurons.
    C. Calcium influx in response to stimuli is increased.
    D. The amount of neurotransmitter released in response to stimuli is reduced.
    E. All of the answers are correct.

18. The neurophysiological process of sensitization is a prolonging of the action potential brought about by which of the following?
    A. slowing of sodium channels opening, resulting in reduced calcium inflow
    B. slowing of potassium channels opening, resulting in reduced calcium inflow
    C. slowing of sodium channels opening resulting in increased calcium inflow
    D. slowing of potassium channels opening, resulting in increased calcium inflow
    E. increase in threshold potential required for cell firing

19. Learning deficiencies in two fruit fly mutations, "dunce" and "rutabaga," share a common underlying cause of:
    A. decreased NMDA
    B. increased NMDA
    C. decreased cAMP
    D. increased cAMP
    E. These two mutations do not share any common basis for their deficiences.

20. Which of the following structural variations is considered to be directly linked to behaviors, individual skills, and memories?
    A. soma size
    B. axon length
    C. number of dendritic spines
    D. thickness of the neural membrane
    E. viscosity of intracellular fluid

Short-Answer Questions

Answer each of the following questions with a brief but complete written answer based on information from your text.

1. Though rare, humans utilize some electrical synapses. Other animals, such as the crayfish, rely much more heavily on such synapses. Briefly describe the primary advantage of electrical synaptic transmission and the primary advantage of chemical neurotransmission.

2. The concept of a quantum of neurotransmitter is important in understanding the magnitude or potential magnitude of a postsynaptic response. Briefly explain what a quantum of neurotransmitter is, and how it relates to postsynaptic neuron response.

3. Briefly describe the four means by which neurotransmitter can be removed from postsynaptic receptor sites to terminate the response.

4. There are four general criteria for identifying a substance as a neurotransmitter. List those four criteria in general terms.

5. One of the most unusual neurotransmitters to be identified is nitric oxide (NO). This neurotransmitter substance is unusual in part because it is a soluble gas (rather than a chemical compound). Briefly describe at least two other ways in which NO is different from chemical neurotransmitters.

6. Briefly describe similarities and differences between ionotropic and metabotropic receptors.

7. When first studying neurotransmitters, researchers believed that one neuron may contain only one neurotransmitter. Considering recent research findings in the area of neurotransmitters, briefly describe the validity of this original hypothesis.

8. The autonomic nervous system uses neurotransmitters to regulate body states during times of arousal and times of relaxation. Briefly explain the primary neurotransmitters that are used by the autonomic nervous system, and how these transmitter substances affect body functions.

9. Eric Kandel and others have done extensive work assessing synaptic changes associated with simple learning in the *Aplysia*. Two types of learning that they have assessed are habituation and sensitization. Briefly describe what is meant by these terms in the sense of learned behaviors.

10. Describe how neural structure is believed to change in association with learning and the formation of new memories. Specifically describe changes that might be associated with habituation and sensitization.

Matching Questions

Complete each of the following matching questions based on information from your text.

1. Match each of the following neurotransmitter substances to the best descriptive feature.

A. Acetylcholine	___ Main excitatory transmitter
B. Glutamate	___ Found at neuromuscular junction
C. GABA	___ Main inhibitory transmitter
D. Opioid peptides	___ Synthesized as a gas
E. Nitric oxide	___ Used for pain reduction

2. Match the following neurotransmitter systems with the disorder in which they are implicated (note that more than one substance has been implicated in some disorders; those disorders have two spaces for matching).

	___, ___ Depression
A. Dopamine	___, ___ Schizophrenia
B. Serotonin	___ Drug abuse
C. Acetylcholine	___ Parkinson's disease
D. Norepinephrine	___ Alzheimer's disease
	___ Manic behavior

3. Match the following research areas with the experimental model that has been most useful for related research.

A. Movement–DA	___ *Aplysia californica* sprayed with water
B. Excitation/inhibition–NE/ACh	___ MPTP-exposed humans
C. Habituation/Sensitization–Ca^{2+}	___ Drosophila with rutabaga mutation
D. Learning–cAMP	___ Heart muscle fibers

4. Identify each of the following as features of either ionotropic (I) or metabotropic (M) receptors.

___ Binding site attached directly to membrane pore
___ Changes cell activity through a series of steps
___ Associated with G-proteins
___ Mediates rapid changes in membrane voltage
___ Structurally similar to voltage-sensitive channel
___ Often utilizes second messenger systems

5. Mark each of the following as being associated with sympathetic (S) or parasympathetic (P) nervous system arousal.

___ Rest-and-digest
___ Heart rate increase
___ Cholinergic neurons
___ Adrenergic neurons
___ Fight-or-flight

Diagrams

1. On the diagram of a synaptic connection below, identify the following: mitochondrion, synaptic vesicle, storage granule, synaptic cleft, presynaptic membrane, postsynaptic membrane, presynaptic terminal, dendritic spine.

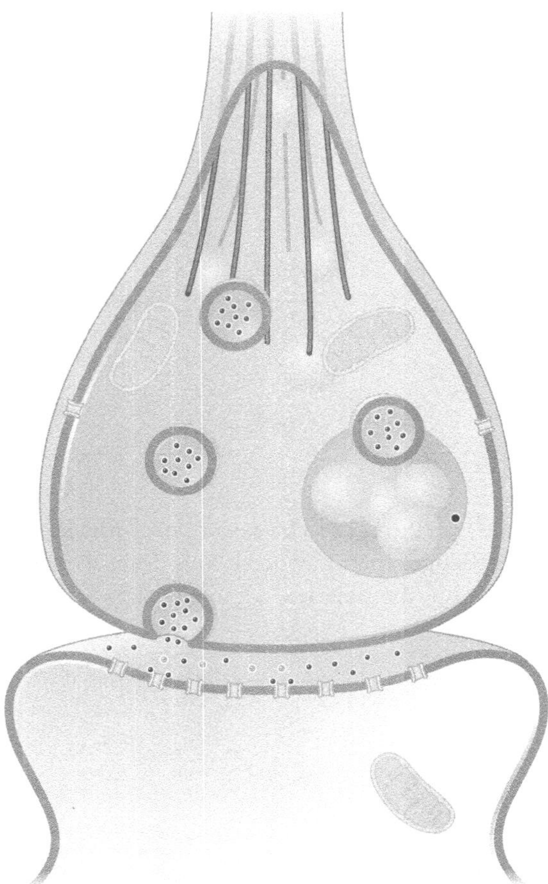

2. Draw a simple diagram to represent each of the following synaptic connections: axodendritic, axosomatic, axoaxonic, axosynaptic, dendrodendritic, axoextracellular, axosecretory.

3. The diagram below depicts the synthesis pathway for the neurotransmitter epinephrine. Fill in the empty boxes.

4. Below is a diagram depicting normal innervation between a sensory and motor neuron. Draw a simple diagram to depict changes in these connections that would be expected with sensitization. Draw a second diagram to depict changes in these connections that would be expected with habituation.

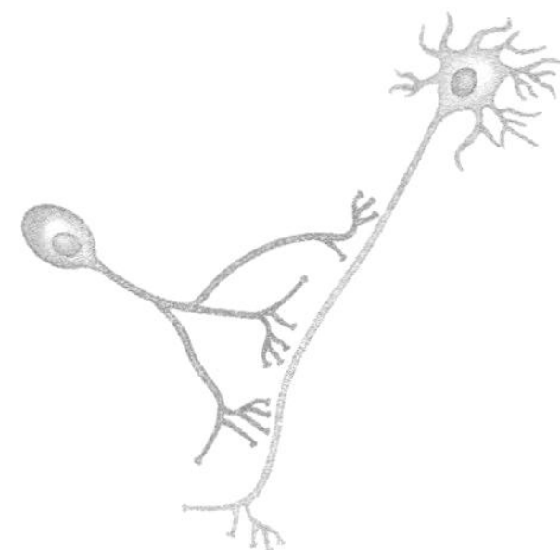

5. Numerous synaptic changes are thought to be associated with learning and memory. Identify at least 4 possible changes that may occur and draw arrows indicating where such changes would be found in the synapse.

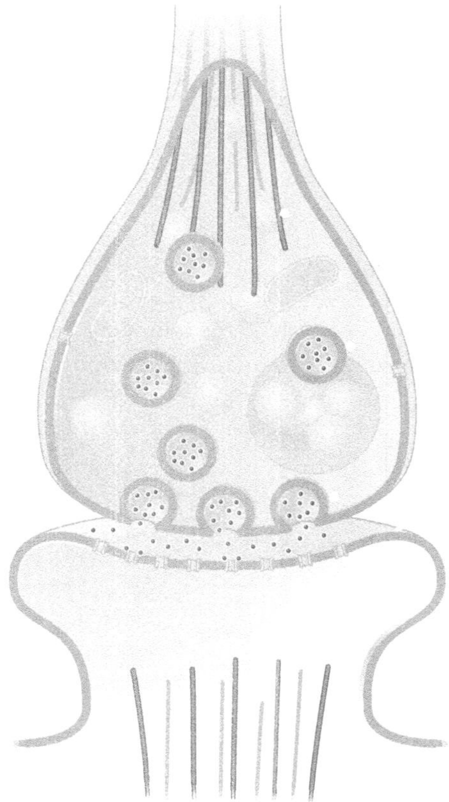

The Web

Consider using the following Web sites for additional information on some of the topics from this chapter:

1. Neurotransmission/Neurotransmitters: www.blackwellpublishing.com/matthews/neurotrans.html
2. Center for the Neural Basis of Cognition: www.cnbc.cmu.edu/
3. Aplysia site including videos: www.brembs.net/learning/aplysia/
4. Parkinson's Disease Foundation Inc.: www.pdf.org/
5. World Parkinson's Disease Association: www.wpda.org/

CROSSWORD PUZZLE

Across

1. Enzymes convert 19 down into this neurotransmitter
6. Acetylcholine is made up of choline and _____.
9. Second word in 21 down; abbr.
10. Very common inhibitory transmitter; abbr.
11. Very common excitatory transmitter
13. One of two enkephalin prefixes
18. Place in England to enjoy a drink and stimulate 18 across receptors
21. Major dopamine pathway; _____ striatal
22. Opposite form of learning from 4 down
24. Endoplasmic reticulum; abbr.
25. Number of classifications of synapses
27. Nicotinic acetylcholine receptor; abbr.
30. Soluble gas messenger; _____ monoxide
31. Large vesicles are called _____ granules.
32. Abbreviated term for "endogenous morphine"

Down

1. 1 across, 15 down, and 19 down
2. First word in 10 across across; abbr.
3. Term for the synaptic space
4. Growing used to a stimulus, also considered a simple form of learning
5. One of two enkephalin prefixes
7. 30 across; abbr.
8. Receptor stimulation may evoke this type of messenger system
12. Term for pump that actively takes neurotransmitters back into terminal
14. Epinephrine; abbr.
15. Enzymes convert 1 across into this neurotransmitter
16. Term for movement from terminal to soma
17. Involuntary movement sometimes seen with L-dopa treatment
19. This neurotransmitter is depleted in Parkinson's disease
20. Detector: Measures autonomic nervous system changes
21. Soluble gas messenger; abbr.
23. Involuntary movements seen in Parkinson's disease
26. Prefix with 32 across
28. 1st neurotransmitter to be identified; abbr.
29. Prefix for region 1 or 3 in the hippocampus

CHAPTER

6 How Do Drugs and Hormones Influence the Brain and Behavior?

CHAPTER SUMMARY

Psychoactive drugs are chemical compounds that act to alter mood, thought processes, or behavior. Such drugs may be used to manage neuropsychological illness or may conversely become *abused substances.* Psychoactive drugs may be administered via numerous routes, including oral ingestion, inhalation, absorption through the skin, or injection into the blood, muscle, or brain. Although oral ingestion is usually the most convenient route, injection and inhalation are generally more effective, bypassing digestive breakdown and other potential barriers.

The effectiveness of a drug is based in part on the route by which it is administered. For example, an effective oral dose of amphetamine is about ten times greater than that required to be effective if inhaled or injected intravenously. In turn, the effective inhaled or injected dose is about ten times greater than that required to be effective if injected into cerebral spinal fluid. There are also individual differences in determining the effective dose of a drug. For example, large people are less sensitive than small people, men are less sensitive than women, and middle-aged people are less sensitive than the elderly.

All drugs (unless injected directly into the brain) must pass through the *blood–brain barrier*, a term for the small capillary and glial network that form *tight junctions* restricting most substances in the bloodstream from entering the brain. For drugs to pass the blood–brain barrier, they must be nonionized fat-soluble small molecules, or they must have a chemical structure resembling substances that are moved through the barrier by an *active transport system* (such as glucose). There are some regions where the blood–brain barrier is quite permissive in allowing substances to pass into the brain; for example, the blood–brain barrier allows passage of hormones to mediate pituitary function. Toxic substances may pass through *area postrema* of the lower brainstem to stimulate a defensive vomiting response.

Drugs have their effect by acting as *agonists* (increasing effectiveness) or *antagonists* (decreasing effectiveness) of neurotransmission at receptor sites. As an example, *black widow spider venom* acts as an acetylcholine agonist by promoting release of this transmitter. The effect is excess activity at the neuromuscular junction that can result in paralysis. *Botulinum toxin* is an antagonist at these same sites. The result of too little stimulation at these receptors is also paralysis. Understanding how a drug affects a synapse is the first step in effective use of the compound. For example, low concentrations of botulinum toxin have potential for use in treating uncontrolled muscle twitches. South American Indians have also used the acetylcholine antagonist *curare* on their arrow tips to induce paralysis in animals they hunt. *Organophosphates* are a potent class of acetylcholine antagonists that bind irreversibly to receptors and have been used as insecticides and in chemical warfare.

Tolerance to drugs may develop in several ways. *Metabolic tolerance* usually results when, after continued exposure, the system begins to produce more enzymes needed to break down a drug. *Cellular tolerance* refers to compensatory changes in cell structure and activity after repeated exposure to a drug. *Learned tolerance* refers to the ability of an individual to exhibit increasingly normal behavior with repeated use of a drug. Conversely, *sensitization* is when an individual shows an increase in responsiveness to successive drug use. Sensitization appears to occur when drugs are used occasionally (rather than continuously, which produces tolerance), and is thought to result from increased neurotransmitter release with each successive use. Also unlike tolerance, sensitization appears to be more selective in the behaviors affected and the environment in which it occurs.

Psychoactive drugs may be classified in a number of ways. Most frequently they are classified based on their most pronounced psychoactive effect. In this regard psychoactive drugs fall into the five broad categories of *sedative-hypnotic antianxiety agents*, *antipsychotic agents*, *antidepressants*, *opiod analgesics*, and *pychotropics* (which include *psychedelic drugs*).

Sedative-hypnotic and antianxiety agents include *alcohol*, *benzodiazepines* (also known as *minor tranquilizers* or antianxiety agents) and *barbiturates*. These drugs all work as agonists at the $GABA_A$ receptor site to reduce neural activity. *Tolerance* is the phenomenon whereby with continued use, increasingly greater doses of these drugs are required to evoke a desired effect. *Cross-tolerance* refers to the fact that chronic use of one drug from this class may result in tolerance to another drug from the same class.

Antipsychotic agents are sometimes classified as first-generation antipsychotics (FGAs) and second-generation antipsychotics (SGAs). FGAs, the first class of effective antipsychotics developed, have primary pharmacological effects of blocking D2 dopamine receptor sites. More recently developed SGAs have only weak antagonist effects at D2 dopamine receptor sites, but have the additional effect of blocking 5-HT2 serotonin receptor sites. These drugs are used to reduce the symptoms of schizophrenia. However, since they are dopamine antagonists (particularly at the *D2 receptor* subtype) they have a side effect of producing symptoms similar to Parkinson's disease. The effectiveness of these drugs in treating schizophrenia was a basis for the development of the *dopamine hypothesis of schizophrenia*, which states that symptoms of this disease result from overactivation of the dopamine system.

Antidepressants include *monoamine oxidase (MAO) inhibitors* and the *tricyclic antidepressants*. MAO inhibitors prevent the breakdown of serotonin and dopamine in the synaptic cleft, acting as an agonist for these neurotransmitters. Similarly, tricyclic antidepressants act as agonists by blocking the reuptake transporter for these neurotransmitters. *Second-generation antidepressants* include *selective serotonin reuptake inhibitors* (*SSRIs*), including fluoxetine (marketed under the name Prozac). Although these drugs are quite successful in reducing symptoms in many patients, about 20 percent of patients do not respond to drug therapy or cannot tolerate the side effects of these treatments. The class of drugs known as *mood stabilizers*, used primarily to treat *bipolar disorder*, includes lithium salt. Another mood stabilizer, valproate, is also used to treat some individuals with epilepsy. These drugs share a feature of reducing the intensity of one pole of bipolar disorder (mania or depression) making the transition to the opposite pole less likely to occur.

Opioid analgesics include *opium*, *morphine*, and *codeine*, all of which are derived from seeds of the opium poppy. All of these drugs are endorphin agonists and act as potent painkillers. *Heroin* is a semisynthetic form of morphine that is more fat-soluble and thus more readily penetrates the blood–brain barrier. *Naloxone* and *nalorphine* are endorphin antagonists and have therapeutic value in treating individuals who have overdosed on endorphin agonists.

Psychotropics include *cocaine* and *amphetamine* and act as potent dopamine agonists. Cocaine, derived from the coca shrub, acts by blocking reuptake of dopamine into the presynaptic terminal. Amphetamine is a synthetic compound that shares the reuptake blocking effect of cocaine, but also has the effect of increasing dopamine release, making it a more potent agonist. Caffeine is a relatively mild stimulant that indirectly increases *cyclic adenosine monophosphate* (*cyclic AMP*, or cAMP), subsequently increasing glucose utilization. *Psychedelic drugs* include *mescaline*, *tetrahydrocannabinol* (*THC*), *lysergic acid diethylamide* (*LSD*), and *psilocybin*. These drugs share a general agonist effect, although the specific neurotransmitter systems they affect may vary.

Uncontrolled alcohol consumption is the most common form of abnormal drug use and offers a useful model for assessing behavioral changes during drug intoxication. Abnormal behavior patterns under alcohol intoxication have been explained using the *disinhibition theory*, which states that cortical function that normally mediates reasoning and higher level cognitive function is depressed by alcohol. The term *alcohol myopia* has also been used to describe how, when under the influence of alcohol, individuals tend to respond much more strongly to cues in their immediate environment, while giving less consideration to distant cues. In both cases, alcohol is thought to reduce normal functioning of brain regions associated with reasoning.

Substance abuse is excessive reliance on and chronic use of a drug. This condition can advance to *drug addiction* and is usually accompanied by physical dependence on the drug. *Withdrawal* is the unpleasant physical condition that accompanies drug disuse. Withdrawal symptoms range from sweating, anxiety, and nausea to convulsions and death, depending on the drug and the extent of dependency. Although a wide range of psychoactive agents are abused, all seem to share the quality of producing *psychomotor activation*, likely through activation of dopamine neurons. It is not clear why humans abuse and become addicted to drugs. The dependency hypothesis suggests that individuals become addicted to drugs in an attempt to avoid unpleasant withdrawal symptoms. However, this does not explain why people return to drug use long after withdrawal symptoms have subsided, or why some drugs that produce few if any unpleasant withdrawal effects are potentially addictive. A more recent hypothesis is that addiction begins through *associative learning* when the *pleasure* of drug intoxication is *classically conditioned* to drug use. Terry Robinson and Kent Berridge have proposed the *wanting-and-liking theory* whereby drug use is initiated by liking the pleasurable effects. As tolerance develops, greater doses are required to increase the liking response. Eventually, very little liking is experienced and the behavior is continued predominantly through wanting because of conditioning to cues associated with drug use. Robinson and Berridge suggest that opioid neurons may be responsible for the liking aspect of drug use, while *mesolimbic dopamine neurons* control wanting behavior.

Research suggests that genetics may predispose some individuals to addiction. For example, if one identical twin abuses alcohol, the other twin has a higher than normal likelihood of also being an alcohol abuser. Children of alcohol-abusing parents who are adopted into homes are also more likely to show alcohol use patterns similar to their biological parents than to non-alcohol-abusing adoptive parents. Finally, individuals who have a genetic predisposition toward risk-taking behaviors also exhibit a higher likelihood of experimenting with drugs than the general population. However, at this time the evidence is based in large part on correlation research, and a genetic abnormality has not been identified that causes addictive behavior.

Some drugs of abuse are known to directly produce neurodegeneration in animal models. For example, administration of MDMA (ecstasy) in doses used by humans results in degeneration of fine serotonergic fibers in rats and monkeys. In other cases, drug addiction may indirectly result in neurodegeneration. For example, chronic alcohol consumption is often accompanied by poor diet and vitamin deficiency that can cause damage to specific brain structures. Other drugs that produce profound psychoactive effects, such as LSD, have not been shown to produce cell loss in animals when administered in doses used by humans.

Hormones are a class of endogenous chemicals that circulate through the bloodstream to target organs. Once reaching their targets, they act in a manner similar to neurotransmitters, altering cell function and ultimately behavior of the organism. Hormones are used to maintain *homeostasis* (such as eating), to regulate *reproductive function* and in response to *stress*. The hypothalamus utilizes *releasing factors* to stimulate the pituitary gland into releasing hormones that then control release of hormones throughout the system. For example, the hypothalamus regulates insulin release from the pancreas, which controls glucose storage and utilization, ultimately mediating eating behavior. The hypothalamus also regulates hormones of the menstrual cycle. Interestingly, changes in ovarian hormones such as estrogen and progesterone during the menstrual cycle in turn appear to affect performance on some motor and cognitive tasks. Stress hormones include *epinephrine*, responsible for rapid effects associated with the adrenaline surge, and *cortisol*, which is utilized for preparing the body for extended periods of stress by decreasing digestion, immune responses and reproductive functioning. Stress responses activated by these hormones occur during “good stress” as well as “bad stress.” Although the response is generally considered adaptive, Sapolsky and his colleagues have recently shown that prolonged and/or extreme stress can result in neurodegeneration, particularly to hippocampal cells.

KEY TERMS

The following is a list of important terms introduced in Chapter 6. Give the definition of each term in the space provided.

Principles of Psychopharmacology

Drug

Psychoactive drugs

Substance abuse

Blood–brain barrier

Endothelial cells

Tight junctions

Active transport systems

Area postrema

Agonists

Antagonists

Black widow spider venom

Botulinum toxin

Curare

Organophosphates

Tolerance

Metabolic tolerance

Cellular tolerance

Learned tolerance

Sensitization

Classification of Psychoactive Drugs

Sedative-hypnotics

Antianxiety agents

Alcohol

Barbiturates

Benzodiazepines

Minor tranquilizers

Tolerance

Cross-tolerance

$GABA_A$

Antipsychotic agents

First-generation antipsychotic (FGA)

Second-generation antipsychotic (SGA)

D_2 receptor

Dopamine hypothesis of schizophrenia

Antidepressants

Monoamine oxidase inhibitors

Tricyclic antidepressants

Second-generation antidepressants

Selective serotonin reuptake inhibitors

Bipolar disorder

Mood stabilizers

Opiate analgeisics

Opium

Codeine

Morphine

Heroin

Nalorphine

Naloxone

Endorphins

Stimulants

Cocaine

Amphetamine

Cyclic AMP

Psychedelic drugs

Tetrahydrocannabinol (THC)

Anandamide

Lysergic acid diethylamide (LSD)

Drugs, Experience, Context, and Genes

Disinhibition theory

Alcohol myopia

Substance abuse

Substance dependence

Addiction

Withdrawal symptom

Psychomotor activation

Dependency hypothesis

Pleasure

Classical conditioning

Incentive sensitization

Mesolimbic dopamine system

Wanting-and-liking theory

Monosodium glutamate (MSG)

MDMA

Hormones

Endocrine glands

Steroid hormone

Peptide hormone

Homeostasis

Gonadal (sex) hormones

Glucocorticoids

Organizational hypothesis

Stress hormones

Homeostatic hormones

Stress response

Epinephrine

Posttraumatic stress disorder (PTSD)

KEY NAMES

The following is a list of important names introduced in Chapter 6. Explain the importance of each person in the space provided.

Sigmund Freud

Terry Robinson

Jill Becker

Kent Berridge

Ian Whishaw

Tara MacDonald

Robert Sapolsky

PRACTICE TEST

Multiple-Choice Questions

Answer each of the following multiple-choice questions with the best possible answer based on information from your text.

1. Which of the following is true of psychoactive drugs?
   A. They may alter mood.
   B. They may alter behavior.
   C. In high doses some may act as toxins in the brain.
   D. They have the potential to become abused substances.
   E. All of the answers are correct.

2. In general, which route of administering drugs produces the most rapid psychoactive effects?
   A. oral
   B. injection into muscle
   C. injection into bloodstream
   D. inhalation
   E. All routes produce effects in about the same amount of time.

3. Which of the following is the most accurate statement regarding the blood–brain barrier?
   A. It does not allow any substances into the brain from the bloodstream.
   B. It is composed of tight junctions formed by endothelial cells.
   C. It is especially effective in keeping lipid-soluble drugs from entering the brain.
   D. It utilizes no active transport of substances, only passive diffusion.
   E. It effectively prevents L-dopa in the bloodstream from entering the brain.

4. The venom from the bite of a black widow spider would likely do which of the following?
   A. kill a human
   B. affect dopamine
   C. act as an agonist
   D. cause muscle twitches in an area around the bite
   E. block release of a neurotransmitter

5. Organophosphates bind irreversibly to acetylcholinesterase. As such they are highly potent agonists at acetylcholine receptor sites. What is a potential use for these compounds?
   A. treating myasthenia gravis
   B. treating Parkinson's disease
   C. reducing muscle twitches
   D. a toxin for chemical warfare
   E. None of the answers is correct.

6. The liver produces enzymes that are used to degrade alcohol. With regular alcohol consumption a liver will begin to produce more of this enzyme than it did prior to alcohol exposure. This phenomenon is termed:
   A. cellular tolerance.
   B. metabolic tolerance.
   C. behavioral tolerance.
   D. sensitization.
   E. intoxication.

7. Tolerance appears to develop with frequent repeated drug use. Sensitization is thought to develop with which of the following?
   A. frequent repeated drug use
   B. frequent repeated drug use followed by long periods of abstinence
   C. initial exposure to the drug
   D. following an episode of drug overdose
   E. occasional use

8. Which of the following is *not* considered a member of the drug classification sedative-hypnotics?
   A. barbiturates
   B. alcohol
   C. Prozac
   D. benzodiazepines
   E. minor tranquilizers

9. Many drugs that reduce anxiety or have minor tranquilizing qualities act as agonists at the $GABA_A$ receptor subtype. What is the ionic change associated with activation of this receptor?
   A. an influx of sodium ions
   B. an influx of potassium ions
   C. an influx of calcium ions
   D. an influx of chloride ions
   E. an efflux of all the ions listed above

10. Antipsychotic agents are effective in reducing the symptoms of what disorder?
    A. depression
    B. mania
    C. Parkinson's disease
    D. chronic pain
    E. None of the answers is correct.

11. Antidepressants are thought to act by improving chemical transmission of serotonin, noradrenaline, histamine, acetylcholine, and dopamine. However, agonists for one of these neurotransmitters have been particularly successful in alleviating symptoms of depression. Select that transmitter from the list below.
    A. serotonin
    B. noradrenaline
    C. histamine
    D. acetylcholine
    E. dopamine

12. The class of drugs known as mood stabilizers:
    A. is not well understood.
    B. includes the salt lithium.
    C. is used to treat bipolar disorder.
    D. includes drugs used to treat epilepsy.
    E. All of the answers are correct.

13. Opium is derived from seeds of the opium poppy. Which of the following is synthesized directly from opium?
    A. heroin
    B. morphine
    C. endorphins
    D. naloxone
    E. nalorphine

14. Cocaine and amphetamine share the feature of blocking dopamine reuptake, making them effective dopamine agonists. What additional feature does amphetamine have at the synapse that makes it an even more potent agonist than cocaine?
    A. It enhances dopamine release from the terminal.
    B. It blocks enzymatic breakdown of dopamine.
    C. It directly stimulates postsynaptic dopamine receptor sites.
    D. It blocks reuptake of serotonin.
    E. All of the answers are correct.

15. Which of the following is *not* classified as a psychedelic drug?
    A. mescaline
    B. tetrahydrocannabinol (THC)
    C. naloxone
    D. lysergic acid diethylamide (LSD)
    E. psilocybin

16. MacDonald and coworkers have proposed that some undesirable behaviors associated with alcohol intoxication result from an inability to attend to distant cues combined with a strong influence of salient cues in the immediate environment. What is the term used by MacDonald to describe this phenomenon?
    A. cue inattention
    B. selective intoxication syndrome
    C. alcohol-induced blindness
    D. alcohol myopia
    E. beer goggles

17. Which of the following is *not* a feature of addiction as defined in your text?
    A. It develops as an advanced state of substance abuse.
    B. It usually is based on use of drugs that inhibit psychomotor activation.
    C. It usually includes physical dependence.
    D. It usually includes development of tolerance.
    E. Withdrawal symptoms are usually experienced if drug use is discontinued.

18. Which of the following is true of how addicts report drug use experience?
    A. Initial drug experience is unpleasant, but pleasure is derived with repeated use.
    B. Initial drug experience is pleasant, but pleasure decreases with repeated use.
    C. Initial drug experience is pleasant, and pleasure is maintained with repeated use.
    D. Initial drug experience is pleasant, and pleasure increases with repeated use.

19. The hypothalamus produces releasing factors to stimulate which gland into releasing hormones?
    A. pineal
    B. pituitary
    C. gonads
    D. adrenal
    E. pancreas

20. Sapolsky contends that the type of stress that stimulates a hormone response is which of the following?
    A. stress associated with fear
    B. stress associated with aggression
    C. stress associated with sadness
    D. stress associated with happy events
    E. stress associated with any of the above

Short-Answer Questions

Answer each of the following questions with a brief but complete written answer based on information from your text.

1. The blood–brain barrier (BBB) may be viewed as both a blessing and a curse. Briefly describe the makeup of the BBB. As a blessing, describe a useful function of the BBB. As a curse, explain how the BBB may hinder pharmacological therapy for some brain disorders.

2. Briefly describe tolerance and sensitization. Give a general explanation of synaptic changes in neural transmission that are thought to underlie these phenomena.

3. The effectiveness of neuroleptics at reducing symptoms of schizophrenia led to development of the dopamine hypothesis of schizophrenia. Briefly describe this hypothesis and how it was formulated from results of neuroleptic treatments.

4. Briefly explain the effect of MAO inhibitors on neural transmission. Also identify the disorder for which these drugs are most frequently prescribed.

5. Early in its history, amphetamine had several potential benefits for which it was used. List at least three reasons amphetamine was used for reasons other than recreation. Also include a statement as to why it is now seldom used therapeutically.

6. Briefly describe the features of disinhibition theory as it applies to alcohol abuse and addiction. Include in your description possible weaknesses of this theory.

7. Briefly define substance abuse, addiction, and withdrawal symptoms. Explain how these three features of drug use are related.

8. Briefly describe differences between men and women in their use of, and response to, psychoactive drugs.

9. There is a body of evidence suggesting that alcoholism, and possibly drug addiction in general, may have a genetic basis. Give at least two examples of research to support this contention. Also explain a weakness in interpreting results from this research.

10. Hormones are generally thought of as mediating sexual behavior. However, a separate class of hormones is utilized in response to stress. Give an example of at least two stress hormones, and explain how these hormones affect behavior.

Matching Questions

Complete each of the following matching questions based on information from your text.

1. Number from 1 to 5 (1 being the fastest, 5 being the slowest) the relative speed with which a drug would have its action when delivered via the following routes of administration.

   __ Oral (eaten)
   __ Topical (applied to the surface of the skin or mucus)
   __ Intravenous (injected into a vein)
   __ Intracranial (injection into the brain)
   __ Inhalation (smoked)

2. Label each of the following actions as agonists or antagonists.

   __________ block presynaptic reuptake of neurotransmitter from synapse
   __________ block enzyme that breaks down neurotransmitter
   __________ block postsynaptic receptor site
   __________ block release of neurotransmitter from presynaptic terminal
   __________ increase effectiveness of neurotransmission
   __________ decrease effectiveness of neurotransmission

3. Match the following drugs to the appropriate feature or characteristic.

   A. Alcohol
   B. Neuroleptics
   C. MAO Inhibitors
   D. Opium
   E. Caffeine

   ___ Block dopamine receptor sites
   ___ Affects $GABA_A$ receptor
   ___ Increases cyclic AMP
   ___ Among the most potent analgesics
   ___ Contributed to development of Prozac

4. Match the following drug class to the appropriate source.

   A. Norepinephrine psychedelics
   B. Tetrahydrocannabinol
   C. Psychomotor stimulant
   D. Opiate analgesics
   E. Sedative-hypnotics antianxiety

   ___ Fermented and distilled
   ___ Coca plant
   ___ Hemp plant
   ___ Poppy seeds
   ___ Peyote cactus sugars

5. Match the following disorders to the appropriate symptoms or characteristics.

   A. Fetal alcohol syndrome
   B. Drug-induced psychosis
   C. Depression

   ___ More common in women than in men
   ___ Poor nutrition may increase symptoms
   ___ Associated with underdeveloped brain
   ___ Sometimes treated with electroconvulsive therapy
   ___ May be co-diagnosed with schizophrenia

Diagrams

1. In several regions, the blood–brain barrier is very permissive (or even lacking), allowing chemicals from the blood to pass easily into the brain. Use the sagittal brain diagram below to identify three regions where chemicals pass readily from the bloodstream into the brain.

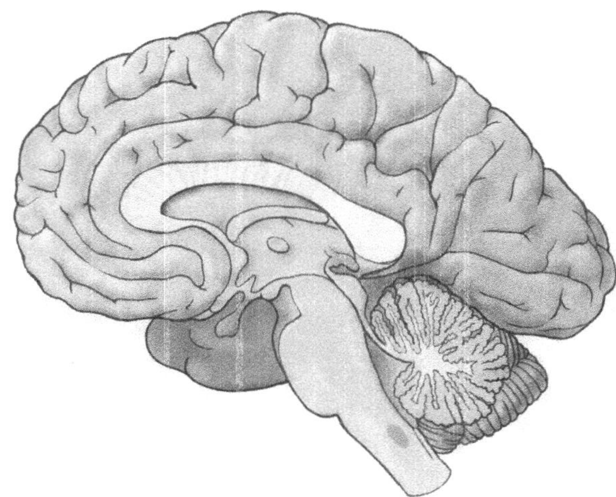

2. Below are three dopamine synapses. The first synapse shows normal function. Depict on the other two synapses the effects of cocaine and a neuroleptic drug.

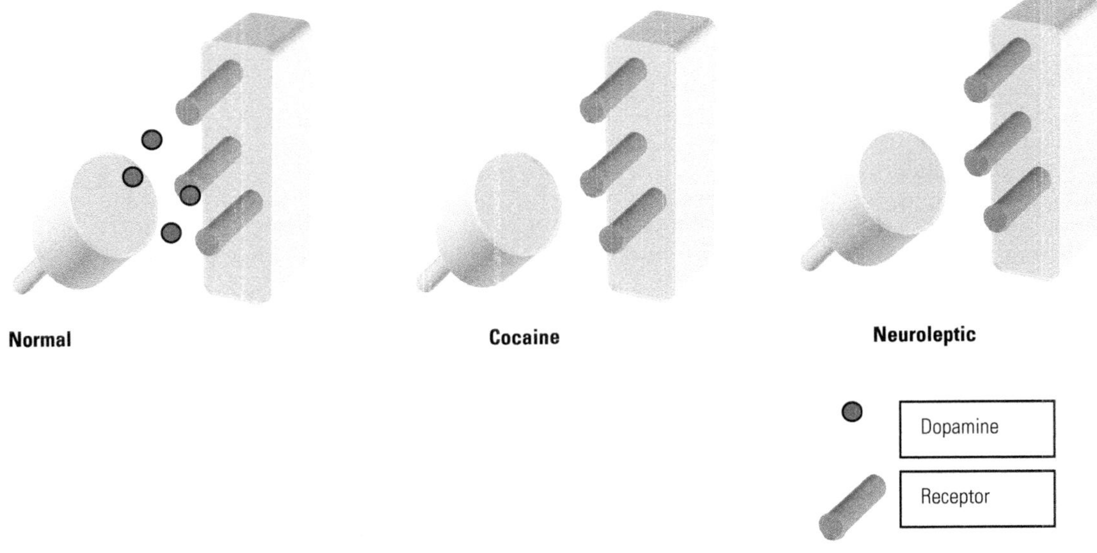

3. Continuous drug use may result in both behavioral tolerance and a physiological tolerance. For example, as individuals consume alcohol over time, tolerance develops for signs of intoxication (behavioral) and for blood alcohol level (physiological). Below is a graph depicting an individual's drinking behavior over a three-week period. Graphically illustrate what would be expected with regards to level of intoxication and blood alcohol level on the other two graphs.

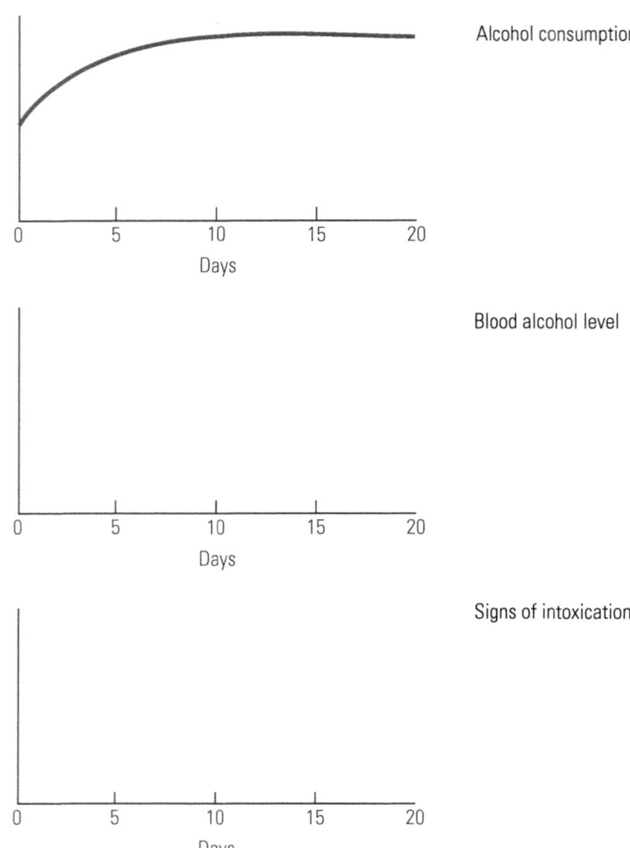

4. Robinson and Berridge have described the process of addiction as involving separate brain mechanisms responsible for both wanting and liking drug use. They further hypothesize that the relative influence of these two mechanisms changes over time with continued drug use. On the graph below, depict with lines the change in relative influence of wanting and liking mechanisms with continued drug use. Assume both mechanisms start at approximately the same level.

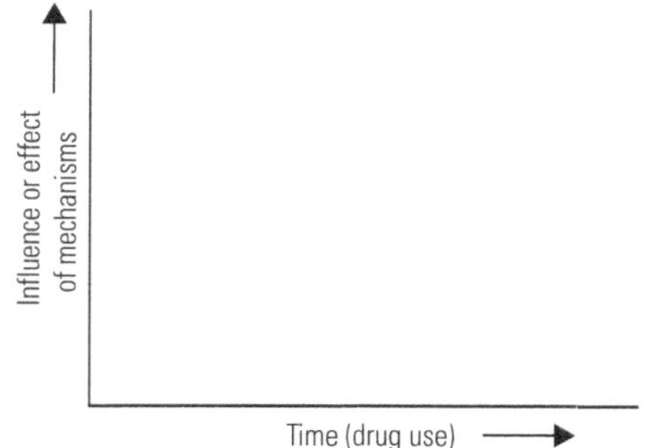

5. Sapolsky and colleagues have suggested a self-perpetuating cycle whereby stress can produce significant brain damage. Complete the diagram below indicating 1) cells affected, 2) change in cortisol secretion, 3) effect of changes in 1) & 2).

The Web

Consider using the following Web sites for additional information on some of the topics from this chapter:

1. National Organization on Fetal Alcohol Syndrome: www.nofas.org/
2. National Institute on Alcohol Abuse and Alcoholism: http://etoh.niaaa.nih.gov/
3. Narcotics Anonymous: www.na.org
4. Partnership for a Drug-Free America: www.drugfree.org/
5. APA division of psychopharmacology: www.apa.org/divisions/div28/index.html

CROSSWORD PUZZLE

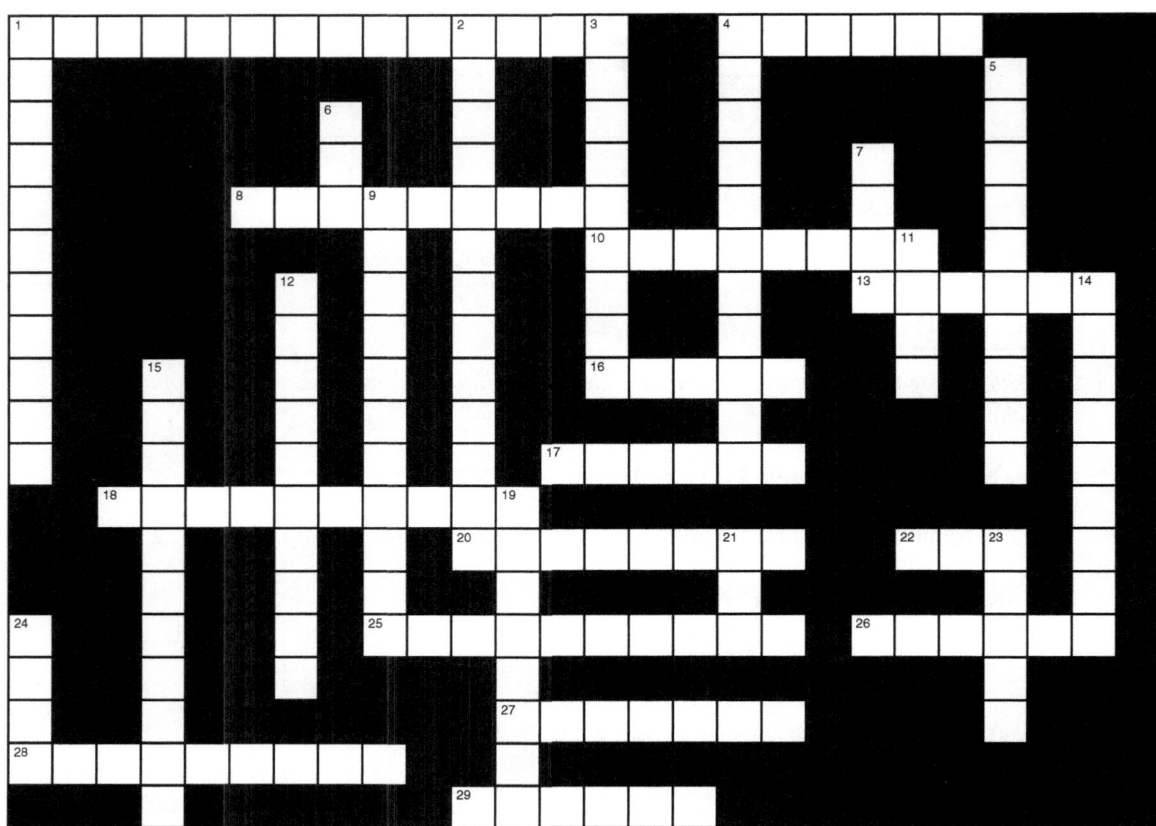

Across

1. Prozac, e.g.
4. One who uses drugs inappropriately or excessively
8. Alcohol and barbiturates
10. Area _____, weak spot in 23 down
13. First word in system from 3 down
16. 7 down is one of these, at serotonin neurons
17. Source of venom that causes ACh release
18. The "G" in FGA and SGA
20. Class of drugs for opium; _____ analgesics
22. Keeps unwanted substances out of the brain; abbr.
25. Codeine and morphine mimic these natural chemicals
26. Extracted from berries, it blocks ACh receptors
27. The "O" in MAO
28. It occurs in a number of stages after repeated drug use
29. Substance synthesized from morphine

Down

1. Benzodiazepine is an example of this class
2. Effect shown by Robinson and Becker after a single injection of amphetamine
3. System that moves glucose into the brain, with 13 across
4. Similar effects, but more potent than cocaine
5. Dopamine system implicated in drug abuse
6. Potent serotonin hallucinogen; abbr.
7. Drug commonly called "ecstasy"; abbr.
9. Type of learning thought to link drug use to pleasure
11. Street name for lysergic acid diethylamide
12. Addicts experience when drug is not taken
14. Gland that releases hormones
15. Term for physiological balance
19. Potent opiate antagonist
21. Drug sensitization researcher and co-author of your textbook: _____ Whishaw.
23. Word for the second letter in 22 across
24. System affected by 8 across; abbr.

CHAPTER

How Do We Study the Brain's Structure and Functions?

CHAPTER SUMMARY

Researchers have long been interested in manipulating behavior by perturbing the brain, and in assessing changes in brain function that are associated with behaviors. Among the first researchers to establish a definitive link between a particular brain structure and a specific function was Paul Broca who, in the late 1800s, noted several cases where brain damage to a certain region of the frontal cortex (now known as *Broca's area*) caused significant speech deficits while having little effect on other behaviors. This, and similar discoveries of the time, marked the genesis of the field of *neuropsychology*, a science dedicated to studying the relationship between brain functions and associated behaviors in humans. A broader perspective on this field of research, that includes the study of both humans and animals, is often referred to as *behavioral neuroscience*. Advancement of both of these fields has relied heavily on advancement in techniques used to quantify behaviors in both humans and animals. These techniques are discussed throughout this chapter.

Neuropsychological testing of memory in humans requires various tasks to assess specific aspects of memory. For example the *Corsi block-tapping* test provides a simple measure of short-term recall used specifically for remembering spatial position of objects. In contrast, *mirror drawing* is used to measure memory of motor skills. Finally, the *recency memory task* is designed to determine a subject's ability not only to recall objects previously observed, but also their accuracy in identifying which objects were observed most recently. Each of these tasks was designed to test specific features of memory that are controlled by distinct brain regions. The use of such an array of tasks to analyze the many characteristics of broader overarching behaviors is common in the field of neuropsychology.

Like the use of multiple tests in humans, animals used in research are often subject to such subdivisions of testing when assessing the relationship between brain structures and behavior. Among the most famous apparatuses for testing memory is the Morris water maze task which can be used to test *place learning* and *matching-to-place learning* with or without a landmark. Each of these tests can be used to evaluate a different feature of learning and memory. Ian Whishaw has developed a paw-reaching task for rats, in which the animals normally exhibit a systematic series of movements to retrieve a food pellet through a small hole in the cage. This relatively simple behavior can be broken down into several distinct features, each of which may be influenced by separate brain structures, making this task potentially quite useful for behavioral neuroscientists.

With techniques designed to measure specific behaviors, the next step in understanding the associated areas responsible for those behaviors is to manipulate specific brain structures. The most basic manipulation is to create a lesion, effectively destroying the structure in question. Karl Lashley was among the first researchers to engage in extensive systematic studies of cortical brain lesions, searching for the locus of memory in rats. His research proved largely ineffective in this area since (as he later learned) the processes of learning and memory are most effectively disrupted by creating lesions of subcortical structures such as the hippocampus. Manipulating subcortical structures, including the process of creating lesions, usually requires the use of a *stereotaxic apparatus*. Such an apparatus allows researchers to accurately insert a probe into deep brain structures where they can produce nonspecific *electrolytic lesions*, or more selective *neurotoxic lesions*.

In contrast to lesions, *brain stimulation* is used to evoke behaviors from specific brain regions by stimulating cells with a low current of electricity. Wilder Penfield was a pioneering scientist in the area of evaluating the effects of cortical stimulation on behavior in awake and alert humans. Rats in which electrodes have been placed in some deep structures will engage in *electrical self-stimulation*, suggesting that the effects of stimulation may evoke a pleasurable sensation. Furthermore, *deep-brain stimulation* of other structures may bring about therapeutic effects for specific brain disorders. *Transcranial magnetic stimulation (TMS)* is a noninvasive technique that allows researchers to evoke some of the beneficial effects of brain stimulation by passing a magnetic field through the head.

Pairing behaviors with specific measures of brain activity is another technique used by neuropsychologists and behavioral neuroscientists. The earliest of these techniques, the electroencephalograph (EEG), was developed in the mid-1900s. The EEG records gross electrical activity in the brain using electrodes placed on the scalp. Using this method, researchers found that the brain exhibits a variety of activity patterns that are indicative of particular states of arousal or even behavioral activities. As such, the EEG has become a useful tool for relating brain activity to behavior in both normal brains, and brains affected by damage or disease. One particularly useful finding was the discovery of *event-related potentials (ERPs)* that are apparent on an EEG when specific stimuli are present. Using ERPs to evaluate brain responses to behaviors and stimuli has the advantage of being noninvasive and relatively inexpensive. Like ERPs and EEGs, *magnetoencephalograms (MEGs)* are generated using a noninvasive apparatus that utilizes a magnetic field passed through the brain as a means of generating data on brain activity.

In the 1970s, the development of *computerized tomography (CT scan)* became the first technique developed specifically to visualize the brain. CT scans use multiple x-ray images to create a three-dimensional image of the brain that clearly distinguishes structures of significantly different densities such as bone, fluid, and tissue. Magnetic resonance imaging (MRI) is a more recent alternative to CT scans that uses a powerful magnetic field to align hydrogen atoms, which vary in number based on the density of tissue. The end result of the MRI procedure is a three-dimensional image that, like the CT scan, differentiates structures based on their density. Taken one step further, *functional MRI (fMRI)* is capable of detecting and producing images of blood flow changes in specific brain regions. Understanding that specific behaviors require increased metabolic activity in associated brain regions, fMRI can show real-time changes in regional metabolism in a person executing a particular behavior.

Positron emission tomography (PET) shares some of the same principles as fMRI. Knowing that increased metabolism indicates increased brain activity, PET techniques employ injections of mildly radioactively labeled compounds that then accumulate in regions where they are utilized in the brain, producing hot spots that can be detected in a device designed to scan for photon emissions from the brain. PET scans can show regional blood flow, metabolism, receptor density, neurotransmitter levels, and many other features of the brain. Though PET produces images that lack the resolution acuity of CT and MRI, this technique allows researchers to investigate a wide range of physiological processes in brain function, making it an extremely valuable research tool.

Another approach to evaluating brain activity associated with behavior is to directly measure chemical changes. In some cases, abnormal behaviors are directly correlated with dysfunctions of brain chemistry. Parkinson's disease, for example, is characterized by slow and labored voluntary movement. These behaviors are directly associated with decreases in dopamine caused by degeneration of dopamine-producing cells in the substantia nigra. Other behavioral changes may likewise be correlated to neurochemical fluctuation. For this reason techniques in monitoring regional changes in brain chemicals can be useful tools for behavioral neuroscientists. *Microdialysis* is a technique whereby a small probe can be implanted into a specific brain region for the purpose of withdrawing samples of brain chemicals. This technique can be used to assess changes in chemicals caused by damage to the brain (as in Parkinson's disease) or to monitor changes in chemicals as they occur during specific behaviors. Like microdialysis, *cerebral voltammetry* can be used to measure levels of extracellular chemicals. This technique, however, uses measures from electrical currents to calculate relative levels of local chemicals, rather than evaluating the chemicals directly by retrieving them from the brain.

The field of behavior genetics is rapidly expanding and as such researchers are now looking for more ways to determine the effects of genes on certain behaviors. In humans, a long-established technique for gaining a gross measure of genetic contributions to a behavior has been to assess *concordance rates* among family members. For these studies, identical twins are particularly useful because they share nearly identical genetic material. Thus, high concordance rates for a behavior seen in identical twins, compared to non-identical twins (or other siblings), typically suggest a genetic contribution to the behavior. Another valuable approach to assessing genetic influences in humans is to evaluate behaviors of children adopted away from their parents and siblings. Similarities seen between siblings raised in separate environments also indicate a genetic contribution to behaviors. Beyond simple genetic contributions, researchers are also now considering how experience can influence the expression of certain traits known to be mediated in part by genetics. *Epigenetics* is the study of how experiences alter gene expression.

Because of the limitations of using humans for studying the direct relationship between brain function and behavior, animals are often used to develop working models. The first question raised, when using animal models for human behaviors, is whether these models are accurate or useful. The answer to this question seems to depend on the behavior in question. In the case of localized brain damage, as seen with a stroke or in Parkinson's disease, animal models are particularly useful. However, it also seems that when carefully selected and evaluated, animal models for more complex disorders may also be highly useful. For example, one line of research using a rat model has produced interesting results related to *attention-deficit/hyperactivity disorder (ADHD)*. One point of agreement among all researchers using animal models is that the welfare of those animals should be at the forefront of consideration when designing and executing an experiment. In some cases, governments have set strict standards for researchers in order to protect the health and well-being of animals used in the lab. Having followed these guidelines is often a mandatory requirement for research results submitted for publication. However, the vast majority of researchers are acutely aware of the fact that animals that are treated with the highest standards tend to be the best subjects for experiments, producing the most accurate and consistent results over time and across experiments.

KEY TERMS

The following is a list of important terms introduced in Chapter 7. Give the definition of each term in the space provided.

Measuring Brain Activity

Functional near-infrared spectroscopy (fNIRS)

Neuropsychology

Broca's area

Behavioral neuroscience

Corsi block-tapping test

Mirror-drawing task

Test of recent memory

Place learning task

Matching-to-place learning task

Landmark-learning task

Brain lesions

Stereotaxic apparatus

Akinesia

Brain stimulation

Electrical self-stimulation

Deep-brain stimulation (DBS)

Transcranial magnetic stimulation (TMS)

Repetitive TMS (rTMS)

Hypokinetic

Hyperkinetic

Measuring the Brain's Electrical Activity

Electroencephalography (EEG)

Event-related potentials (ERP)

Magnetoencephalography (MEG)

Extracellular recording

Intracellular recording

Place cells

Static Imaging Techniques: CT and MRI

X-ray

CT scan

Magnetic resonance imaging (MRI)

Dynamic Brain Imaging

Functional magnetic resonance imaging (fMRI)

Positron emission tomography (PET)

Near-infrared spectroscopy (NIRS)

Optical tomography

Chemical and Genetic Measures of Brain and Behavior

Microdialysis

Cerebral voltammetry

Concordance rate

Brain-derived neurotrophic factor (BDNF)

COMT gene

Epigenetics

Using Animals in Brain–Behavior Research

Attention-deficit/hyperactivity disorder (ADHD)

Kyoto SHR rat

KEY NAMES

The following is a list of important names introduced in Chapter 7. Explain the importance of each person in the space provided.

Paul Broca

Richard Morris

Ian Whishaw

Karl Lashley

Wilder Penfield

Hans Berger

PRACTICE TEST

Multiple-Choice Questions

Answer each of the following multiple-choice questions with the best possible answer based on information from your text.

1. The Corsi block-tapping test provides a measure of which of the following?
   A. short-term recall of spatial positions
   B. long-term recall of spatial positions
   C. fine motor skills
   D. gross motor skills
   E. All of the answers are correct.

2. The maze developed by Richard Morris tests elements of memory in rats that must do which of the following?
   A. avoid electrical shocks
   B. run through a series of tunnels
   C. swim in a vat of water
   D. climb across intertwined ropes
   E. press a sequence of levers for rewards

3. Which of the following is not a specific movement seen in rats executing the skilled reaching task developed by Ian Whishaw?
   A. aim the paw
   B. reach over the food
   C. flex the paw digits
   D. grasp the food
   E. withdraw and move food to the mouth

4. Which of the following is an apparatus that allows researchers to manipulate subcortical structures in the brain, such as making a lesion in a rat brain?
   A. bregma
   B. stereotaxic
   C. EEG
   D. MEG
   E. microdialysis

5. Deep-brain stimulation is being used in some human trials as an experimental therapy for which of the following?
   A. Alzheimer's disease
   B. alcoholism
   C. Parkinson's disease
   D. multiple sclerosis
   E. color blindness

6. Using EEG, early researchers discovered which of the following is true?
   A. EEG waves do not vary as individuals engage in different behaviors.
   B. EEG activity is silent when an individual is sleeping.
   C. Some EEG waves show a rhythmical pattern.
   D. The highest amplitude waves in an EEG are seen when a person is aroused or excited.
   E. All of the answers are correct.

7. Which of the following methods of assessing brain activity would be considered non-invasive?
   A. EEG
   B. ERP
   C. MEG
   D. All of the answers are correct.
   E. None of the answers is correct.

8. CT scans utilize which of the following to produce three-dimensional images?
   A. magnetic fields
   B. electrical fields
   C. X-rays
   D. radioactivity
   E. None of the answers is correct.

9. The "f" in fMRI suggests which of the following capabilites?
   A. It can show brain responses as they occur in real time.
   B. It can show multiple perspectives of the brain.
   C. It can compare brain images over a developmental period of months or years.
   D. It can make predictions about how the brain will behave in the future.
   E. It is a fast form of MRI, taking about half the amount of time required for a regular MRI.

10. Which of the following is *not* true of PET scans?
    A. They create an image based on the measurement of photons emitted from the brain.
    B. They require that a person be injected with a radioactive compound.
    C. They can generate images based on relative levels of blood flow.
    D. They can generate images based on relative numbers of neurotransmitter receptor sites.
    E. They can generate images based on relative density of tissue.

11. NIRS tomography works on the principle that an object can be reconstructed by gathering which of the following when it is transmitted through an object?
    A. light
    B. magnetic fields
    C. radioactive particles
    D. sound waves
    E. electricity

12. One advantage of using NIRS is that it is relatively easy to hook subjects up for evaluation at different points throughout their lifetime. Which of the following is one disadvantage?
    A. It is expensive.
    B. It is painful.
    C. It is difficult to measure activity in the cortex using this technique.
    D. It is difficult to measure activity in structures below the cortex using this technique.
    E. This technique requires subjects to remain very still for many hours at a time.

13. Microdialysis is a method of collecting brain chemicals by employing which of the following physical forces?
    A. simple diffusion
    B. electricity
    C. magnetic fields
    D. suction
    E. thermal energy

14. In terms of evaluating brain function, the technique of cerebral voltammetry is most similar to which of the following?
    A. NIRS
    B. fMRI
    C. CT scan
    D. in vivo microdialysis
    E. PET Scan

15. Concordance rates for a genetic trait would likely be lowest for which of the following pairs?
    A. mother and daughter
    B. brother and sister
    C. father and daughter
    D. father and son
    E. husband and wife

16. Caspi and colleagues (2005) found that carriers of the VAL allele were far more likely to develop psychosis if they:
    A. had abusive parents.
    B. had a sibling with psychosis.
    C. used cannabis during adolescence.
    D. had a traumatic childhood.
    E. consumed a high-fat diet.

17. Changes in gene expression can result from which of the following experiential factors?
    A. exposure to drugs
    B. chronic stress
    C. contraction of a disease
    D. experiencing traumatic events
    E. All of the answers are correct.

18. Animal models showing that methylphenidate can improve performance of rats on tests of attention processes are particularly useful for the study of which of the following disorders?
    A. Alzheimer's disease
    B. Parkinson's disease
    C. stroke
    D. ADHD
    E. depression

19. At the time this text was published, how many organizations made up the Canadian Council on Animal Care that oversees laboratory animal care and use?
    A. 2
    B. 10
    C. 20
    D. 100
    E. 200

20. U.S. regulations specify that researchers consider alternatives to procedures that may cause which of the following?
    A. momentary slight pain to an animal
    B. more than momentary slight pain to an animal
    C. the birth offspring from experimental animals
    D. boredom of an animal
    E. obesity in an animal

Short-Answer Questions

Answer each of the following questions with a brief but complete written answer based on information from your text.

1. List the three neuropsychological tests of memory discussed in this chapter. Briefly describe what type of memory is tested with each.

2. Briefly explain why Karl Lashley was unable to find the location of memory in the brain, in spite of spending 30 years making and analyzing cortical lesions.

3. Briefly summarize the differences between electrolytic lesions and neurotoxic lesions. Describe a situation in which one of these types of lesions might be preferred over the other.

4. Analysis of ERPs has gained popularity among researchers in recent years. Briefly describe advantages of using ERPs compared to other devices for measuring brain activity associated with certain behaviors or in response to certain environmental stimuli.

5. Compare and contrast the techniques of CT and MRI as they are used for brain imaging.

6. Explain how fMRI works in terms of being able to detect changes in brain activity associated with specific behaviors.

7. Briefly describe how microdialysis is performed and the type of data that may be generated using this technique.

8. Briefly explain what concordance rates are and how they are used in assessing genetic influences on behavior, particularly among related individuals.

9. Briefly describe the field of epigenetics and how both human and animal models can be used for research in this field.

10. Briefly summarize the four guidelines used by the Canadian Counsel on Animal Care when they review the protocols for proposed studies that will utilize animals.

Matching Questions

Complete each of the following matching questions based on information from your text.

1. Match the following tests with the behaviors they are designed to measure.

A. Associated with the Morris water maze	_____ Corsi block-tapping test
B. Tests for motor skill recall	_____ Mirror-drawing task
C. Tests for short-term recall of spatial position	_____ Recency memory task
D. Long series of cards, each with 2 stimulus items	_____ Matching-to-place learning
E. Studied in depth by the authors of your text	_____ Paw-reaching behavior

2. Identify each of the following as either a technique that requires invasive (I) or noninvasive (NI) procedures.

_____ Transcranial magnetic stimulation
_____ Deep-brain stimulation
_____ Electrical lesions
_____ Event-related potential recording (ERP)
_____ Optical tomography

3. Match the following imaging techniques with their appropriate characteristic.

fMRI CT EEG PET NIRS

_____ Uses X-ray
_____ Uses a magnetic field
_____ Uses radioactive compounds
_____ Measures electrical currents
_____ Measures light transmission

4. Match the following animal model to the appropriate human disorder.

A. Parkinson's disease	_____ Kyoto SHR rat
B. Amnesia	_____ Interrupted blood supply to the brain
C. ADHD	_____ Neurotoxic lesions of dopamine pathways
D. Stroke	_____ Infant rats deprived of maternal contact
E. Childhood abuse	_____ Electrolytic lesion of the hippocampus

5. Match the name of the researcher with the technique he helped develop.

A. EEG	_____ Whishaw
B. Cortical stimulation	_____ Morris
C. Cortical lesions	_____ Berger
D. Paw reaching	_____ Lashley
E. Water maze	_____ Penfield
F. Evaluation of language centers in the brain	_____ Broca

Diagrams

1. Identify the correct task being used in each of the Morris water-maze paradigms depicted in the three diagrams.

A. Place-learning task
B. Matching-to-place task
C. Landmark-learning task

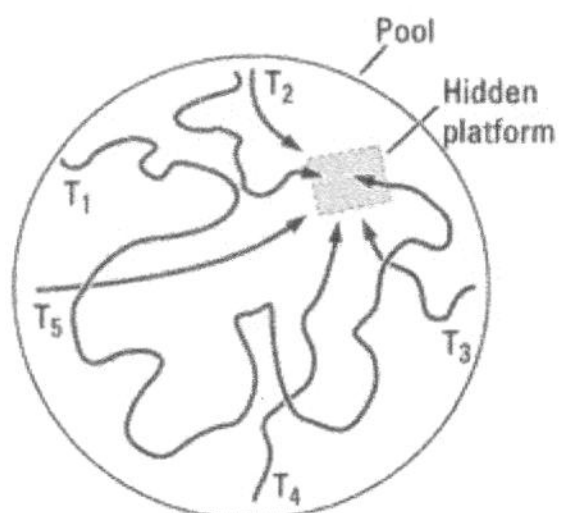

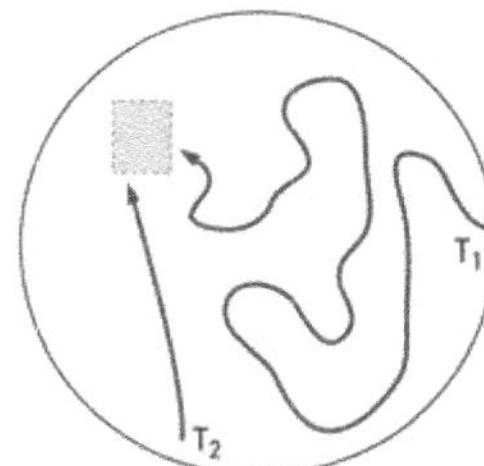

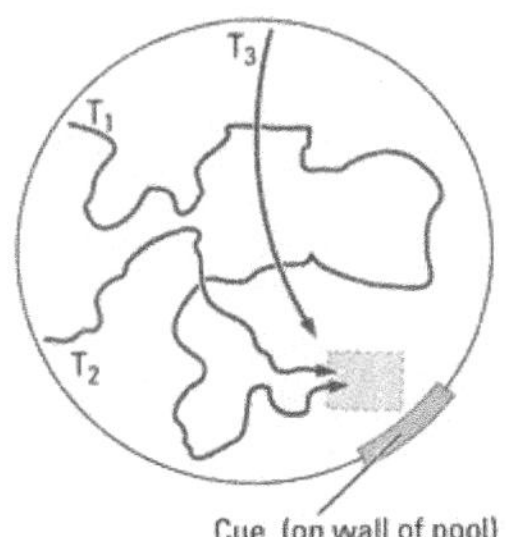

2. Draw representations of what EEG recordings from an individual in each of the following states might look like:

Awake and excited

Drowsy

Relaxed with eyes closed

3. Identify the technique used to produce each of the two images below. Note the perspective from which each of these images is being viewed (dorsal, lateral, anterior). Label the lateral ventricles in both images.

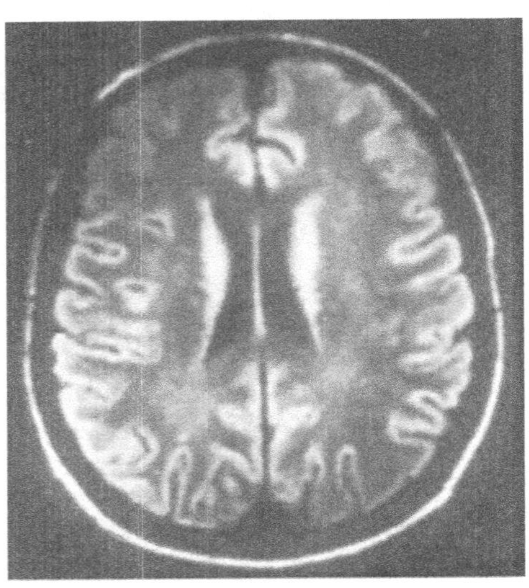

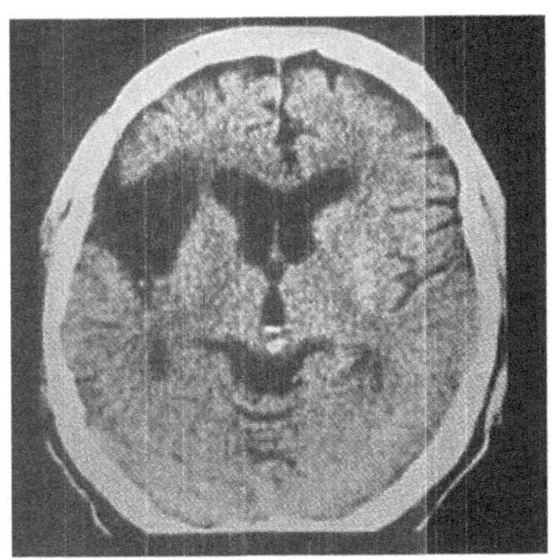

4. Label the brain recording or brain-imaging technique being depicted in each of the following diagrams.

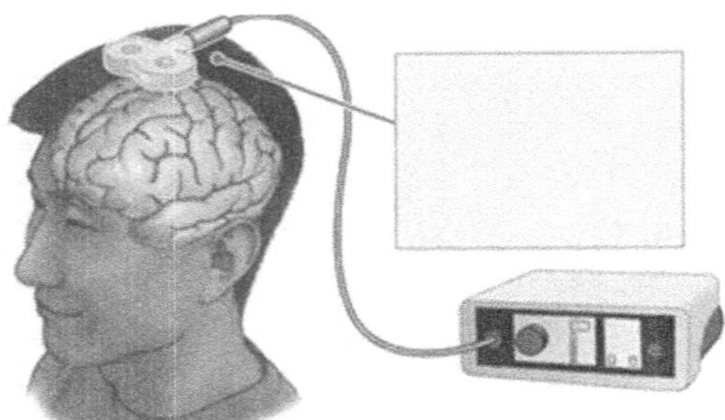

A.

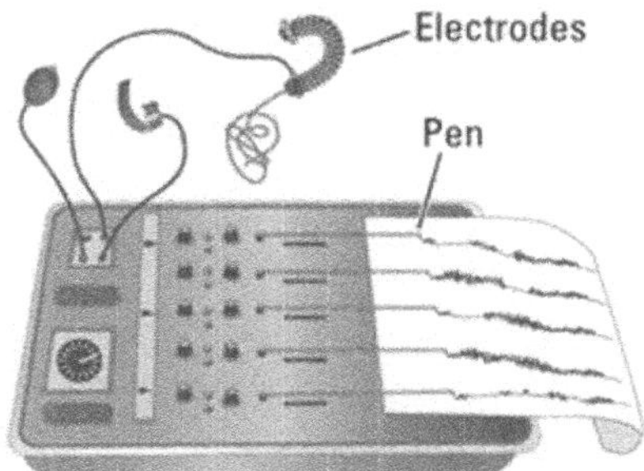
Electrodes
Pen

B.

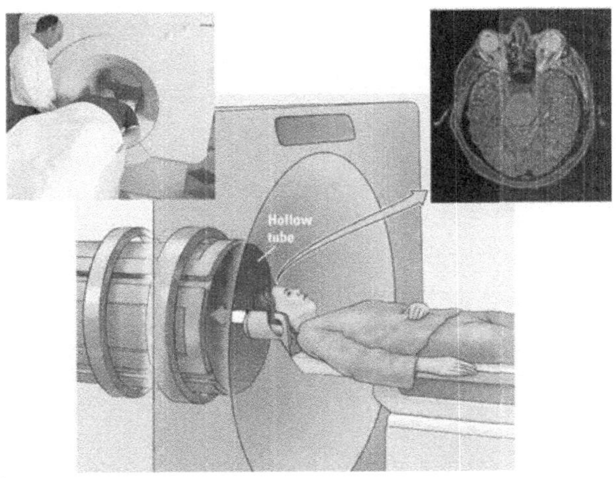
Hollow
tube

C.

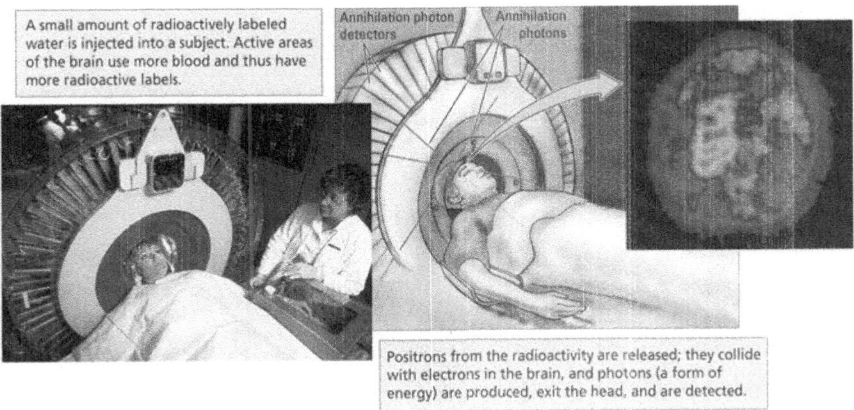
A small amount of radioactively labeled water is injected into a subject. Active areas of the brain use more blood and thus have more radioactive labels.
Annihilation photon detectors
Annihilation photons
Positrons from the radioactivity are released; they collide with electrons in the brain, and photons (a form of energy) are produced, exit the head, and are detected.

D.

5. Add labeled arrows to the diagram below that show each of the following:

   A. Movement of molecules through the semipermeable cannula membrane
   B. Flow of fluid into the brain
   C. Flow of fluid containing the collected sample out of the brain

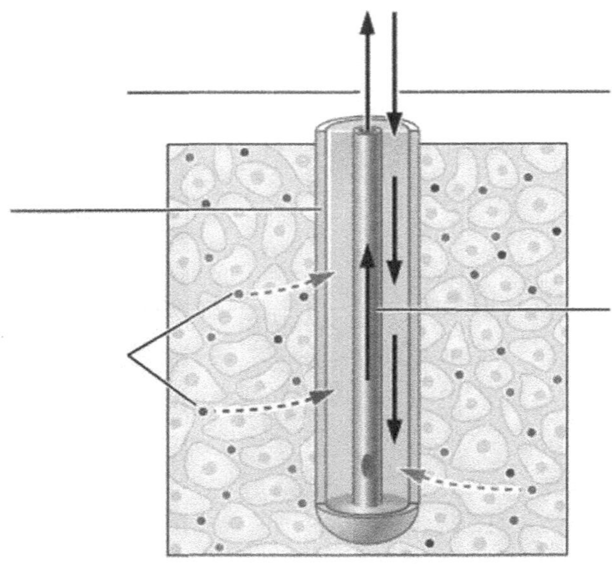

The Web

Consider using the following Web sites for additional information on some of the topics from this chapter:

1. Oxford Centre for Functional MRI of the Brain: www.fmrib.ox.ac.uk

2. General information on radiology techniques: www.radiologyinfo.org/

3. Search "Morris Water Maze" using: www.youtube.com

4. Imaging techniques: www.nmr.mgh.harvard.edu/martinos/research/technologies.php

5. Deep brain stimulation site:
   http://biomed.brown.edu/Courses/BI108/BI108_2008_Groups/group07/index.html

CROSSWORD PUZZLE

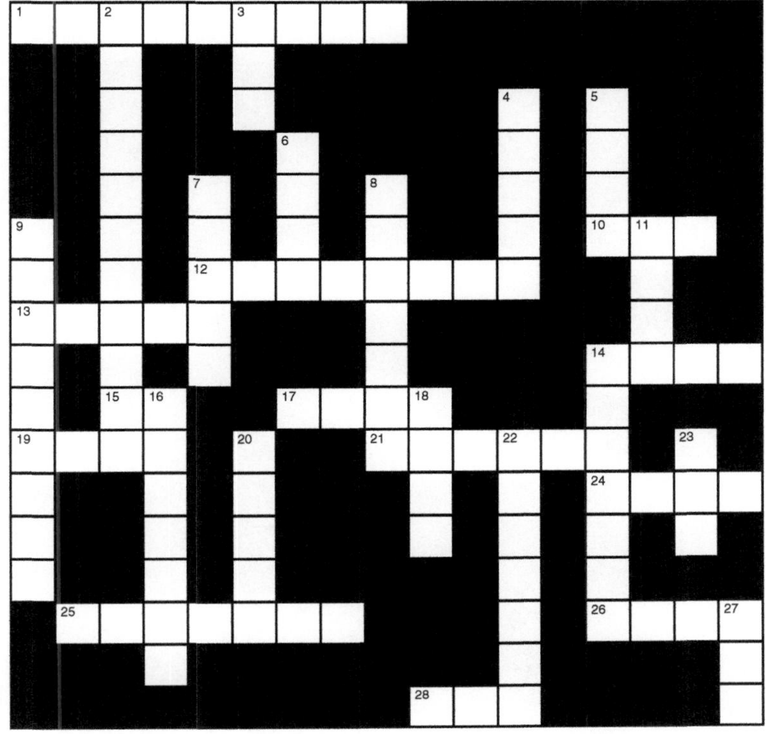

Across

1. May be extracellular or intracellular
10. Brief change in an EEG signal; abbr.
12. The "P" in PET
13. Cell that codes for spatial location; _____ cell
14. The "D" in DBS
15. Imaging technique that uses X-rays; abbr.
17. PET _____ or 17 across
19. Language center "Broca's _____"
21. To destroy a brain region
24. Disease modeled in Kyoto rats; abbr.
25. Too much movement, with 7 down
26. Gene that links cannabis and psychosis
28. Hans Berger was a pioneer in this technique

Down

2. Measure of trait frequency, often for twins
3. Invasive technique with therapeutic value; abbr.
4. Organ of interest for neuroscientists
5. DNA segment that carries vital information
6. Noninvasive technique that gathers light; abbr.
7. Too much movement, with 25 across
8. One form of tomography
9. Invasive device; stereotaxic _____
11. Word used with 2 down, or with "interest"
14. _____ brain imaging; e.g., fMRI or PET
16. Corsi test requires this finger movement
18. The "N" in NRIS
20. The "E" in ERPs
22. Noninvasive _____ techniques
23. Specifically, 24 across refers to Kyoto _____ rats
27. Magnetic stimulation of the brain; abbr.

CHAPTER

8 How Does the Nervous System Develop and Adapt?

CHAPTER SUMMARY

To understand the influence of brain development on emergence of behaviors, several strategies may be used. One method is to analyze structural development and correlate these changes to the emergence of behaviors. A second method is to first assess behavioral changes, and then draw conclusions about neural development based on these observations. Finally, factors that affect both brain and behavior may be assessed. Such factors include the influence of hormones, toxins, and injury during development.

The study of brain development has roots in philosophical discussions nearly 2000 years old. Early philosophers subscribed to the concept of *preformation*, or the idea that fetal development constituted a time when the organism was miniaturized, and that development was merely an enlargement process. Preformation was the dominant theory until the mid-nineteenth century, when detailed analysis of embryos began to reveal too many anatomical differences during gestation to support this theory. Today we know that brain development begins in a curled sheet of cells called the *neural tube*. *Neural stem cells* proliferate in this tube, which later forms the ventricular system for the brain. Stem cells give rise to *progenitor cells* that may further divide and produce nonproliferating *neuroblasts* and *glioblasts*. During development, neural tissue does not take on the appearance of a brain until about 100 days, and the human cortex surface features of gyri and sulci do not become apparent until nearly seven months.

It is believed that cell development into specific types of cells is mediated in large part by genetic codes contained in the DNA. However, chemical *neurotrophic factors* may also be involved. Two examples of such chemicals are *epidermal growth factor* (*EGF*), which stimulates stem cells to produce progenitor cells, and *basic fibroblast growth factor* (*bFGF* or *FGF-2*), which stimulates progenitor cells to produce neuroblasts.

Neurogenesis (the process of forming neurons) is complete at around 4.5 months of gestation. Cell migration from the neural tube to appropriate destinations begins shortly after cell production begins, and continues for about six weeks after neurogenesis ceases. Migration of most neurons is guided by *radial glial cells*, while a small number of cells appear to be guided by chemical signals. Migration guides cells first to deep brain structures, with layers of cells added on until the outermost cortical layers are formed at the end of development. Once in place, cells begin to mature or differentiate. This process includes relatively rapid axon development, with a slower though more extensive process of growth and *arborization* of dendrites. Axonal development is led by a *growth cone*, which then divides into several *filopodia* (or finger-like projections), which eventually form terminals. Growth cones are attracted to or repelled from sites by *tropic molecules*. Only one group of such molecules, called *netrins*, has been identified, but researchers are quite certain more exist. *Cell-adhesion molecules* (*CAM*) are a class of molecules that allow growth cones to adhere to a surface to guide growth or to form a connection.

The nervous system has a propensity to vastly overproduce both cells and neural connections during development, after which time nonfunctional connections are pruned and nonessential neurons degenerate. This process is sometimes referred to as neural Darwinism. The most inactive neurons are most susceptible to cell death, and it is believed that lack of cell activity results in reduced neurotrophic factor and ultimately a genetically programmed death termed *apoptosis*. Cells likely to survive are those that are highly active, especially clusters of cells that fire simultaneously, forming neural circuits for sensory and motor systems.

Once neurons are established, glial development finalizes myelin sheath production around established axons. In fact, myelin formation has been correlated to development of behavioral processes after birth. One example of this is motor coordination that develops in synchrony with myelination of axons in the motor cortex, located in the frontal lobe. Language development also closely correlates to neural development of cortical neurons near Broca's area, located in this same general region. Complex cognitive development has been described in detail beginning with observations of *Jean Piaget*. Piaget and others have described a series of four major cognitive development stages seen in all normally developing children. These stages occur at roughly the same time that *growth spurts* appear in brain development, suggesting that expansion of neural tissue underlies the development of increasingly complex cognitive strategies. Animal models have shown that similar growth spurts appear in monkeys and are accompanied by increasing ability in task-solving strategies.

Brain plasticity refers to the ability of the brain to change throughout life in response to experiences. Such experiences include external stimuli as well as internal stimuli, such as hormones, toxins, and genetic aberrations. Donald Hebb was among the first to show that developing animals exposed to a stimulus-enriched environment developed greater brain mass and neural complexity than animals raised in an environment deprived of such stimuli. From this work it was proposed that increasing neural stimulation during development can reduce synaptic pruning. It has further been proposed that early exposure to language, music, and other culturally influenced stimuli results in long-lasting familiarity with those stimuli. Such plasticity allows an infinite number of possible neural connections modified by experiences during development, generated from a single genetic blueprint.

Particular *critical periods* of development have been identified during which time establishment of neural connections can lead to lasting behaviors. The demonstration by Konrad Lorenz of *imprinting* in baby goslings serves as a prototypical example. Imprinting occurs during a critical period, after which it cannot be established. In cases where imprinting does not occur, or occurs with an inappropriate model, birds exhibit abnormal sexual and social behaviors in adulthood suggesting a permanent change in neural structure and function. Other abnormal experiences encountered during critical periods of neural development (such as deprivation of social contact) have been shown to have long-lasting and sometimes devastating effects in both animals and humans. Internal experiences such as hormones also have their effect during a critical period. For example, the presence of testosterone during development masculinizes neural development, leading to male-pattern behavior later in life. Similarly, females exposed to testosterone during their critical period exhibit some male-pattern behaviors later in life.

Regarding brain injury during development, it appears the second half of the gestational period is a particularly vulnerable time. During this period, after neurogenesis is complete and migration is occurring, damage can produce severe cognitive and motor deficits. *Spina bifida* is a condition in which the back portion of the neural tube does not completely close, resulting in incomplete formation of the spinal cord and severe motor disturbance. *Anencephaly* occurs when the front portion of the neural tube does not completely close. In this case, brain development is severely affected and survival of the infant is generally limited to only a few days. In cases of intellectual disabilities, it appears that dendritic arborization is impaired, resulting in abnormal neural connections. Similar malformation in neural connections may underlie a host of neurological disorders that become apparent later in life, including seizures, dyslexia, and schizophrenia and other mood disorders. Although abnormalities in neural plasticity appear to result in numerous disorders, this same feature of the nervous system serves a very adaptive function of allowing "normal" development in the face of a wide range of internal and external factors.

KEY TERMS

The following is a list of important terms introduced in Chapter 8. Give the definition of each term in the space provided.

Neurobiology of Development

Preformation

Neural plate

Neural tube

Sexual dimorphism

Neural stem cells

Subventricular zone

Progenitor cells

Neuroblasts

Glioblasts

Neurotrophic factors

Epidermal growth factor

Basic fibroblast growth factor

Neurogenesis

Migration

Differentiation

Radial glial cells

Dendritic arborization

Growth cones

Filopodia

Cell-adhesion molecules (CAMs)

Tropic molecules

Netrins

Synaptic pruning

Neural Darwinism

Apoptosis

Behavior and Development

Sensorimotor stage

Preoperational stage

Concrete operational stage

Formal operational stage

Object permanence

Conservation of liquid volume

Growth spurts

Nonmatching-to-sample task

Development and Environment

Brain plasticity

Chemoaffinity hypothesis

Amblyopia

Critical period

Imprinting

Masculinization

Spina bifida

Anencephaly

Phenylketonuria (PKU)

Down syndrome

Fetal alcohol syndrome (FAS)

Rubella

Cerebral palsy

Kwashiorkor

Schizophrenia

KEY NAMES

The following is a list of important names introduced in Chapter 8. Explain the importance of each person in the space provided.

Jean Piaget

Donald O. Hebb

Konrad Lorenz

Harry Harlow

PRACTICE TEST

Multiple-Choice Questions

Answer each of the following multiple-choice questions with the best possible answer based on information from your text.

1. In assessing the relationship between brain and behavior by examining how structural development correlates to behavior, we would anticipate that growth of brain structure associated with language would:
   A. precede development of any language skills.
   B. occur after development of basic language skills.
   C. occur after development of sophisticated language skills.
   D. parallel some aspects of language skills.
   E. not be related in any way to development of language skills.

2. The early and predominant theory of embryo development known as *preformation* was eventually dismissed because of which of the following findings?
   A. Human embryos did not look like human adults.
   B. Human embryos closely resembled embryos of other species.
   C. Human embryos have a tail.
   D. All of the answers are correct.
   E. None of the answers is correct.

3. The neural tube is sometimes thought of as the nursery in which neural cells proliferate. What happens to this neural tube as the brain develops into adult form?
   A. It is absorbed into brain tissue.
   B. It becomes the basal ganglia.
   C. It becomes the cerebral cortex.
   D. It becomes the vertebrae surrounding the spinal cord.
   E. None of the answers is correct.

4 Which of the following is the term used for cells lining the neural tube?
A. neurons
B. neural stem cells
C. progenitor cells
D. neuroblasts
E. glioblasts

5. Which of the following is the primary function of neurotrophic factors?
A. guide axon growth
B. prune nonfunctional axons
C. stimulate cell production
D. speed neural signals
E. block the action of hormones

6. Neurogenesis is:
A. the pruning of little-used axonal connections.
B. the final stage of brain development.
C. the myelination of axons.
D. the process of forming neurons.
E. another term for apoptosis.

7. Which of the following is true of the process cells undergo when migrating to their appropriate regions of the brain?
A. Most follow chemical signals; a small number utilize radial glial cells.
B. Most follow radial glial cells; a small number utilize chemical signals.
C. Approximately equal numbers utilize radial glial cells and chemical signals.
D. Most migrate randomly; a small number use both radial glial cells and chemical signals.
E. Most use both radial glial cells and chemical signals; a small number migrate randomly.

8. The process of neural development forming the cortical layers appears to follow which of the following sequences?
A. posterior forms first, then anterior
B. anterior forms first, then posterior
C. outer structures form first, then inner structures
D. inner structures form first, then outer structures
E. development varies considerably between individuals

9. Which of the following is true of dendritic arborization?
A. It is a relatively slow process of pruning dendrites.
B. It is a relatively slow process of forming dendritic branches.
C. It is a relatively fast process of pruning dendrites.
D. It is a relatively fast process of forming dendritic branches.
E. It is a process that is thought to be unaffected in Down syndrome.

10. During development, growth cones may send out fingerlike projections known as:
A. radial glial cells.
B. netrins.
C. tropic molecules.
D. dendritic branches.
E. filopodia.

11. Which of the following is true of myelination?
    A. It is not well correlated to maturation of cerebral structures.
    B. Cells cannot function until they are myelinated.
    C. All myelination is generally completed about the time of birth.
    D. Areas that control the highest level of functioning are thought to be myelinated last.
    E. All of the answers are true.

12. Which of the following is *not* true of language development?
    A. It is mediated solely by development of controlled movements of the mouth.
    B. It is controlled in large part by structures in the cerebral cortex.
    C. Areas controlling language increase dendritic density dramatically between 15 and 24 months.
    D. About 1 percent of all children with normal intelligence show marked delays in speech development.
    E. Language acquisition is largely complete by age 12.

13. Piaget noted distinct stages of cognitive development in children that correlate with growth spurts seen in neural development. How many stages of development did Piaget propose?
    A. four
    B. eight
    C. twelve
    D. sixteen
    E. twenty

14. Which of the following is true of animals raised in a stimulus-enriched environment compared with animals raised in an impoverished environment?
    A. Enriched-environment rats perform better in maze-learning tasks.
    B. Enriched-environment rats have larger neurons.
    C. Enriched-environment rats have a greater number of synaptic connections.
    D. Enriched-environment rats have more and larger astrocytes.
    E. All of the answers are correct.

15. In the 1960s, Roger Sperry astutely hypothesized that developing axons are guided to specifically target sites by which of the following?
    A. other axons already established at the target site
    B. electrical signals produced at the target site
    C. enhanced blood flow to the target site
    D. chemicals produced in, or released from, the target site
    E. None of the answers is correct.

16. Most of the Romanian orphans studied by Rutter exhibited dramatic improvement (to near normal measures) in which of the following within two years of being adopted?
    A. height and weight
    B. head circumference
    C. intelligence
    D. All of the answers are correct.
    E. None of the answers is correct.

17. Which of the following is the most accurate statement regarding the influence of hormones on neural plasticity?
    A. Hormones do not influence neural plasticity.
    B. Hormones influence plasticity only during fetal development.
    C. Hormones influence plasticity through childhood.
    D. Hormones influence plasticity through adolescence.
    E. Hormones influence plasticity throughout an entire lifetime.

18. Which of the following is the most accurate statement regarding the influence of hormones on neural development?
    A. Hormones do not influence neural development.
    B. Hormones act alone to influence neural development.
    C. Hormones act to reverse the effects of environmental influences on neural development.
    D. Hormones may mediate the effects of environmental influences on neural development.
    E. Hormones affect neural development only in males.

19. Which of the following is *not* a similarity between spina bifida and anencephaly?
    A. both result in death soon after the infant is born
    B. both have a genetic basis
    C. both result from malformation of the neural tube
    D. both are disorders of fetal neural development
    E. both are considered nontreatable conditions

20. Which of the following is a common feature of children with intellectual disabilities?
    A. elongated axons
    B. abnormal dendritic growth
    C. shortened axons
    D. small cell bodies
    E. abnormally low number of neurons

Short-Answer Questions

Answer each of the following questions with a brief but complete written answer based on information from your text.

1. For more than a thousand years the theory of *preformation* dominated research in fetal development and consequently neural and behavioral development. Briefly define preformation, and explain why this theory was eventually abandoned.

2. Neurogenesis is completed within the first half of a full-term pregnancy. During the second half of the pregnancy neurons undergo the process of pruning and forming functional connections. During which period is the fetus particularly vulnerable to injury or trauma? Briefly explain your answer.

3. Both trophic molecules and tropic molecules are discussed in this chapter. The terms are so similar they are easy to confuse. Briefly describe the function of each of these molecules that are essential for normal neural development.

4. The term *neural Darwinism* is used when describing brain cell development. Briefly describe what is meant by this term.

5. Growth spurts in neural development seen in young children correlate nicely with changes seen in the capacity for cognitive function. Growth spurts, however, are not simply periods of cell proliferation. In fact, few if any new neural cells are added to the brain after the first 4.5 months of gestation. What then accounts for neural growth spurts in children?

6. In one of the more interesting methodological approaches to assessing brain plasticity, Donald Hebb allowed a group of young laboratory rats to grow up in the kitchen of his home. What was his rationale for this and what were his findings?

7. Roger Sperry proposed a theory that formed the basis for the *chemoaffinity hypothesis* used to explain neural connections during development. Briefly explain the chemoaffinity hypothesis.

8. Explain the concept of a critical period of neural development. Use the example of birds imprinting in your explanation.

9. Briefly describe spina bifida as it relates to neural development. Compare this disorder with anencephaly.

10. Plasticity is often used to describe a response to brain injury. However, plasticity is a part of normal development and normal brain function. Give an example of how plasticity can be thought of in terms of normal brain development and function.

Matching Questions

Complete each of the following matching questions based on information from your text.

1. Indicate from first (1) to last (5) the order of development of the following structures and neurons.

   ___ Progenitor cells
   ___ Neural groove
   ___ Neural stem cells
   ___ Neuroblasts and glioblasts
   ___ Neural tube

2. Match the following substances involved in development with their appropriate feature or description.

A. Epidermal growth factor (EGF)	___ Stimulates production of progenitor cells
B. Basic fibroblast growth factor (bFGF)	___ Guide growth cones to the cell
C. Radial glial cells	___ Guide cell migration from neural tube
D. Netrins	___ Stimulates production of neuroblasts
E. Cell-adhesion molecules	___ Provide adhesive surface for guiding cells

3. Match the following terms with the most appropriate definition or description.

A. Neurogenesis	___ Axonal projections
B. Arborization	___ Dendritic branching
C. Filopodia	___ Programmed cell death
D. Apoptosis	___ Cell proliferation
E. Neural Darwinism	___ Process of eliminating neurons

4. Number from earliest (1) to latest (4) Piaget's stages of behavioral development. Then draw an arrow from each stage to the appropriate characteristic(s) of that stage.

A. Conservation, mathematical transformations	___ Formal operational
B. Pretend play, language development	___ Sensorimotor
C. Object permanence, stranger anxiety	___ Concrete operational
D. Abstract logic	___ Preoperational

5. Match the following disorders of neural development with the appropriate feature or description.

A. Amblyopia	___ Lazy eye syndrome
B. Impoverished environment	___ Serious motor problems
C. Spina bifida	___ Abnormal intellect and social behaviors
D. Anencephaly	___ Genetic abnormality
E. Down syndrome	___ Fatal soon after birth

Diagrams

1. On the figure below, identify the approximate location of the following regions: forebrain, midbrain, hindbrain, spinal cord, neural tube.

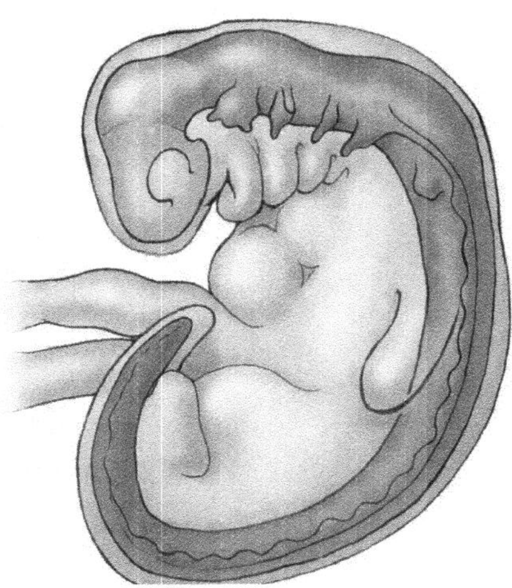

2. Label each of the four cell types seen during development.

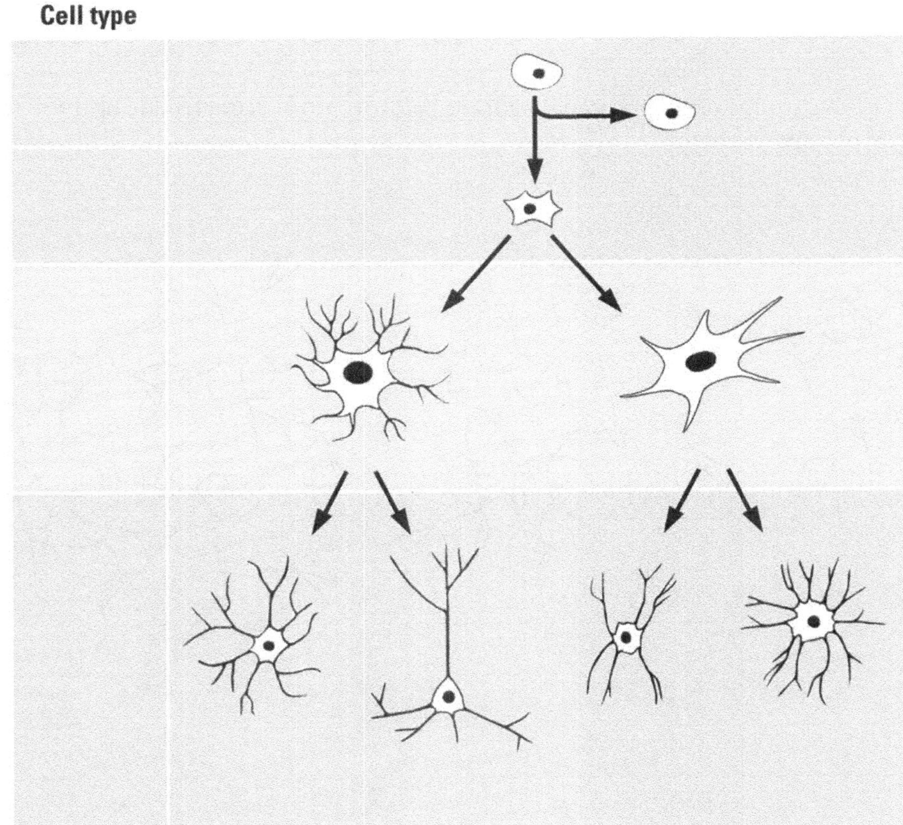

3. Draw below three cortical neurons. Assume Neuron A comes from a rat raised in a dark isolated cage, Neuron B comes from a rat raised with one other rat in a normal caged environment, and Neuron C comes from a rat raised with 10 siblings in a garbage dump.

Neuron A | Neuron B | Neuron C

4. Below are two neurons, one representative of a normal individual, the other representative of an individual with an intellectual disability. Draw appropriate dendritic arbors on the two neurons.

5. Which of the groups of hippocampal neurons below is most characteristic of those seen in schizophrenic patients?

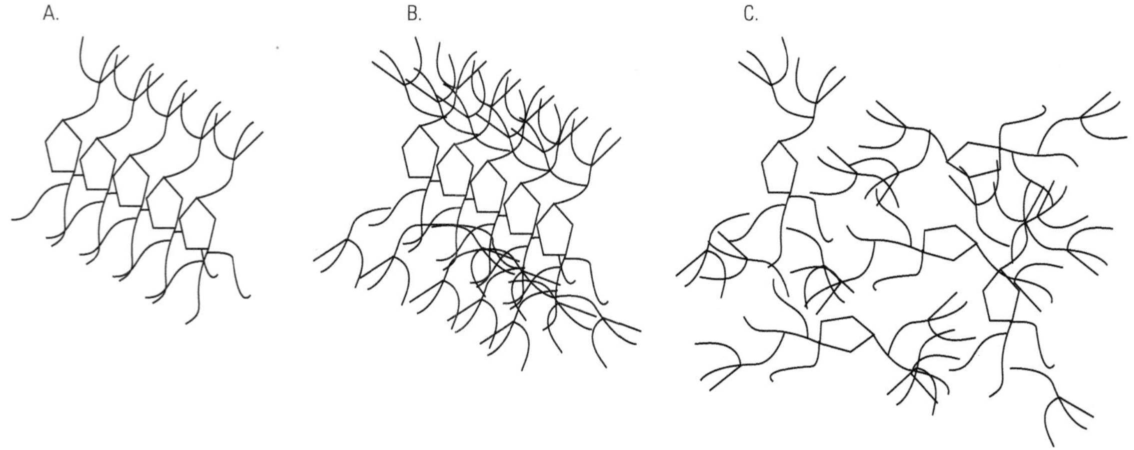

The Web

Consider using the following Web sites for additional information on some of the topics from this chapter:

1. International Society for Prenatal Diagnosis: www.ispdhome.org/
2. United Cerebral Palsy: www.ucp.org/
3. Prental Health and Development:
   www.mayoclinic.com/health/prenatal-care/PR00112
4. National Association for Down Syndrome: www.nads.org
5. Spina Bifida Association: www.spinabifidaassociation.org

CROSSWORD PUZZLE

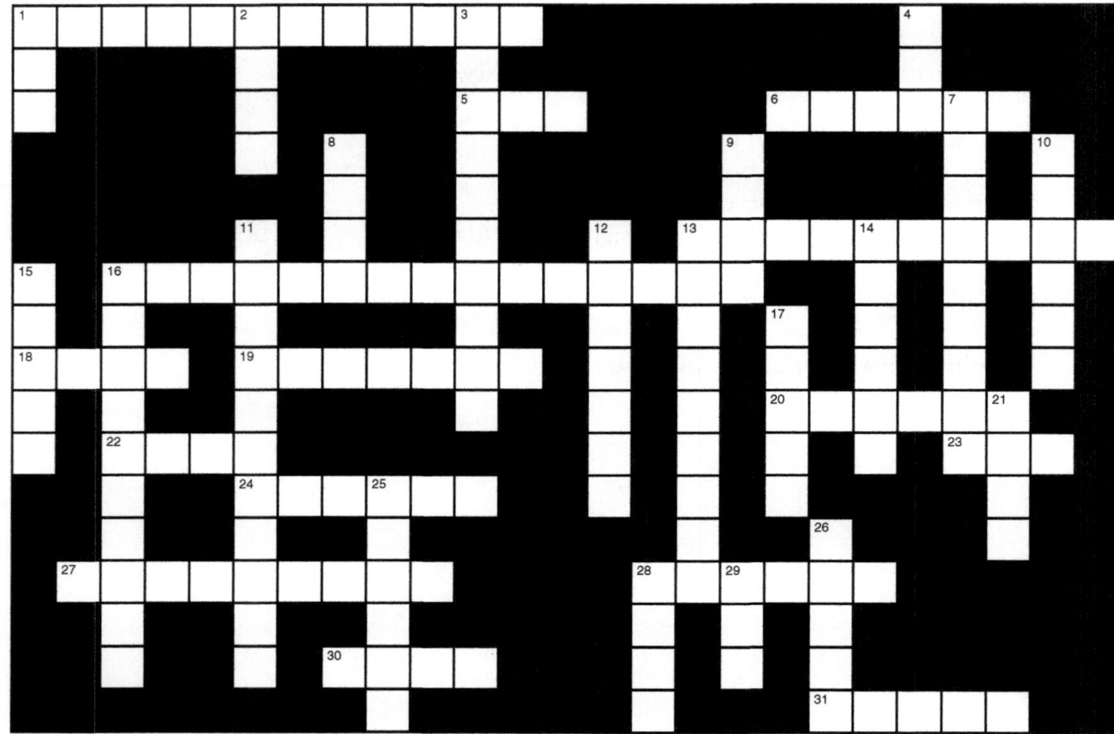

Across

1. The birth of neurons, literally
5. 5 across; abbr.
6. With 31 across, when back of neural tube fails to close during development
13. An axon "foot" extended out during development
16. Genetic abnormality of metabolism
18. Glia do this around axons to form a sheath
19. A class of tropic molecules, meaning "to guide"
20. Signal molecules are also called _____ molecules
22. Form taken after groove
23. 21 down, singular; abbr.
24. He studied childhood development through human observation
27. What cells do during development, and birds do in the spring and fall
28. He showed monkeys raised in social isolation developed abnormally
30. Progenitor cells are derived from this type of cell
31. See 6 across

Down

1. Prefix with natal
2. Cells do this as they mature; grass and flowers do it as well
3. Lorenz showed this phenomenon with ducks
4. Divisions of 17 down development; _____mesters
7. Type of branching that is abnormal in individuals with intellectual difficulties
8. Neuron or glial
9. They guide most axon growth, then form the myelin sheath
10. Growth _____ in the brain coincide nicely with stages of cognitive development
11. When the front of the neural tube fails to close during development
12. German measles, may cause birth defects
13. Like 13 across, except "feet"
14. Time when brain development is most sensitive; critical _____
15. Chromosome abnormality causes this syndrome
16. Reference to the brain's ability to change throughout life
17. Development before birth
21. Cell-adhesion molecules; abbr
25. Guides axon growth; with 26 down
26. Second word in 25 down
28. He let rats live in his kitchen
29. Common animal for studying development

CHAPTER

9 How Do We Sense, Perceive, and See the World?

CHAPTER SUMMARY

As we begin discussion of sensory systems it should be noted that our version of the world, or reality as we know it, is a creation of our brain based on sensory information collected from the environment. Sensory receptors are, in simple terms, neurons that are specialized to convert energy from the environment into neural energy. Environmental energy might consist of photons of light (visual system), sound waves (auditory system), volatile chemicals (olfactory system), and so on. Sensory systems are highly organized, allowing us to discern numerous features of the energy input from our environment. In general, receptor density is positively correlated with the level of sensitivity. This concept is particularly well illustrated by the amount of neural tissue associated with somatosensory input. Topographic mapping of the sensory cortex reveals a high density of receptors associated with hands and face, relative to the receptors that receive input from less sensitive areas of the body, such as the legs and the trunk. Once sensory input reaches the brain, it must be perceived. Perception is the process of assigning meaning to the neural input associated with environmental energy.

Vision is considered the "primary" sensory system for humans because of our reliance on this system. As such, much of sensory-system research (and this book) has been dedicated to understanding vision. Simply stated, vision is a system designed to capture electromagnetic energy, transform that energy into a neural signal, and then interpret the neural signal. Human visual perception is dependent on undertaking this process within a relatively small range (from approximately 400 to 700 nanometers) of electromagnetic energy waves. Beyond the range of our visual perception are ultraviolet waves (too small for us to interpret) and infrared waves (too large for us to interpret).

The structure of the eye is designed to capture light energy. Light passes into the eye through a *cornea* (outer covering) and a pupil. The pupil constricts and dilates, based on movement of the *iris*, in response to high or low light conditions, respectively. The *lens* focuses light energy onto the *fovea*, an area of the *retina* at the back of the eye that contains the greatest density of receptor cells designed for acute vision. It is also worth noting that images are inverted in this process. In other words, our eyes see the world upside down, but our brain is capable of reversing this effect so we perceive our environment as it truly is (right-side up). Visual information leaves the eye via a bundle of neurons through the *optic disc*, a region of the retina that contains no receptors. The optic disc is also known as the *blind spot* because, with no receptor cells, there is no interpretation for the portion of our visual field that falls on this region.

Photoreceptors may be categorized as *rods* or *cones*. Rods, which make up the vast majority of photoreceptors, are long slender cells that are sensitive to dim light and are concentrated in areas outside of the fovea. Cones are cone-shaped cells that are densely packed into the fovea and are used for color vision and for acute visual processing. In general terms, humans have three subtypes of cones that are commonly called "blue," "green," and "red," based on their maximum response to wavelengths of light interpreted as these colors. Some women appear to possess two subtypes of the red cone, potentially increasing to some degree the spectrum of color they can interpret from the environment.

After photoreceptors have transformed light energy into a neural signal, that neural signal is then sent via bipolar cells to retinal ganglion cells (RGCs) and then to the brain. RGCs that receive input primarily from rods are called *magnocellular cells* (or *M cells*), while those that receive input primarily from cones are called *parvocellular cells* (or *P cells*). In the process of cell transmission from receptors, to bipolars, to RGCs, the signal may also be modified by *horizontal* and *amacrine cells* found in the retina.

Fiber tracts from both eyes merge at the optic chiasm on the ventral surface of the brain. Here half the fibers from each eye cross to the opposite hemisphere. Input from the left side of our visual field (as seen by both eyes) goes to the right hemisphere of the brain, while input from the right visual field goes to the left hemisphere. Beyond the chiasm two distinct pathways are formed. All of the projections from P ganglion cells and a few from M cells form the *geniculostriate system*, which sends information first to the lateral geniculate nucleus (LGN) of the thalamus and then on to the *primary and secondary visual cortex* of the occipital lobe. The remaining M cells form the *tectopulvinar system*, which sends information to the *pulvinar* region of the thalamus.

Margaret Wong-Riley and colleagues have identified regions within the visual cortex of the occipital lobe that appear to interpret color (called *blobs*) and regions that interpret motion (*interblobs*). This finding suggests that we have independent systems for different aspects of visual perception that are highly integrated, allowing us to distinguish numerous features of a single object simultaneously. Perception of objects occurs in our *visual field*, which is basically the portion of the environment that we can see. Individual rods and cones have a *receptive field*, the portion of the visual field from which they receive information. As rods and cones converge on a ganglion cell, they form the receptive field for that ganglion. As ganglion cells converge on a visual cortex cell, they similarly form a receptive field that represents the part of the retina to which they are connected. In this way, signals from ganglion cells to cortical regions can produce a *topographical representation* of the entire visual field, allowing us to interpret location of any object we see. It is also worth noting that a particularly large volume of brain tissue is devoted to the interpretation of information coming from the part of our visual field that strikes our fovea.

Intracellular recordings taken from cells in the visual pathways have increased our understanding of how the brain perceives shape. Researchers have found that cells may be either excited or inhibited by light, depending on its location in the receptive field. An *on-center* cell, for example, increases firing when light strikes the center of its receptive field, and decreases firing when light is moved to the periphery of this field. *Off-center cells* show the opposite response to light location. Such cells respond particularly well to edges of light, generating a comparison of light and dark known as *luminance contrast*. By emphasizing the edges of objects, the visual system is capable of perceiving shapes based on light contrast. Information about light contrast interpreted by ganglion cells is sent to *simple*, *complex*, and *hypercomplex cells* of the visual cortex. Simple cells respond vigorously to a stationary bar of light. Complex cells respond best to a moving bar of light. Hypercomplex cells also respond best to a moving bar of light, but have a strong inhibitory area at one end of the receptive field. All of these cells share the feature of responding to luminance contrast along the edge of a stimulus.

Some visual information is processed in area TE of the temporal lobe. This region appears to be responsible for interpreting particularly complex stimuli, including facial recognition and recognition of the same complex object from different orientations (a phenomenon known as *stimulus equivalence*). Cells in this region appear to integrate characteristics such as orientation, size, color, and texture.

Color vision is explained with two separate but integrated theories. The *trichromatic theory* states that color perception is achieved by weighing a ratio of activity from the three separate types of cones, each of which responds greatest to a different wavelength of light. Color perception is achieved by summing the input from each of these types of cells. According to this theory, lacking one of the three types of cones would result in impaired color vision, which is precisely what happens in some types of human color blindness. Furthermore, lacking two types of cones results in an inability to compare ratios and a complete loss of color perception. The *opponent-process theory* is based on the observation that there appear to be four natural colors: red, green, blue, and yellow. Furthermore, red and green, and blue and yellow, appear to oppose each other in the visual system, as demonstrated by the phenomenon of afterimages. This phenomenon is explained at the level of ganglion cells that exhibit on-center and off-center properties in response to opposing wavelengths.

Injury to the visual system beyond the eye results in a variety of blindness disorders. *Homonymous hemianopia* occurs when damage occurs along the visual pathway in one hemisphere and results in loss of perception of an entire visual field. *Scotomas* are blind spots in a visual field produced by small lesions of the occipital lobe. Such blind spots generally go unnoticed (much like our natural blind spot) because involuntary eye movements *(nystagmus)* allow us to process all of the visual field with intact receptors very quickly. *Visual-form agnosia* is an inability to recognize objects, although the ability to manipulate objects remains intact. Another form of agnosia is an inability to see objects when they are in motion. Both types of agnosias result from damage to the ventral stream of visual information, affecting the ability to determine "what" the object is, but not the ability to interact with the object. *Optic ataxia*, on the other hand, is an ability to recognize stimuli but an inability to incorporate appropriate motor patterns for interacting with objects. This deficit results from damage to the dorsal stream of visual information. Visual-form agnosia and optic ataxia illustrate that our sensory systems may utilize numerous strategies, each with different neural pathways, to interpret information from our environment.

KEY TERMS

The following is a list of important terms introduced in Chapter 9. Give the definition of each term in the space provided.

Nature of Sensation and Perception

Receptive field

Optic flow

Auditory flow

Topographic map

Sensation

Structure of the Eye

Cornea

Lens

Iris

Sclera

Retina

Pupil

Blind spot

Fovea

Optic disc

Papilloedema

Photoreceptors

Rods

Cones

Retinal Neurons

Bipolar cells

Horizontal

Amacrine cells

Retinal ganglion cells (RGC)

Magnocellular

Parvocellular

Visual Pathways

Optic chiasm

Geniculostriate system

Striate cortex

Superior colliculus

Pulvinar

Tectopulvinar system

Ventral stream

Dorsal stream

Lateral geniculate nuclei (LGN)

Cortical columns

The Occipital Cortex

Primary visual cortex

Extrastriate cortex

Secondary visual cortex

Blobs

Interblobs

Location in the Visual World

Visual field

Receptive field

Topographic map

Corpus callosum

Neural Activity

On-center cells

Off-center cells

Luminance contrast

Orientation detectors

Simple cells

Complex cells

Hypercomplex cells

Ocular dominance columns

Seeing Color

Trichromatic theory

Opponent-process theory

Color constancy

Injury to the Visual Pathway to the Cortex

Homonymous hemianopia

Quadrantanopia

Scotomas

Visual-form agnosia

Optic ataxia

PRACTICE TEST

Multiple-Choice Questions

Answer each of the following multiple-choice questions with the best possible answer based on information from your text.

1. A visual aura has been reported by some individuals to precede which of the following?
   A. psychotic episodes
   B. sleep
   C. migraine headache
   D. auditory hallucinations
   E. childbirth

2. The distinction between sensory systems may become blurred in a phenomenon called synesthesia where:
   A. vision becomes blurred.
   B. a person becomes tone deaf.
   C. colors are not distinguished.
   D. a person may "hear" visual stimuli.
   E. sensory systems are dulled by drug use.

3. What is the approximate range of wavelengths of electromagnetic energy that a human can perceive as light?
   A. 1–400 meters
   B. 1–400 nanometers
   C. 400–700 meters
   D. 400–700 nanometers
   E. The human visual system does not perceive electromagnetic energy.

4. Approximately 50 percent of the population is affected by myopia. Which of the following is the primary cause of this visual disorder?
   A. lesions of the optic tract
   B. lesions of the occipital lobe
   C. lack of cones
   D. an eyeball that is too long
   E. The cause of myopia is unknown.

5. The blind spot occurs in a region at the back of the eye that:
   A. contains no rods.
   B. contains no cones.
   C. contains nerve fibers leaving the eye.
   D. contains blood vessels entering the eye.
   E. All of the answers are correct.

6. The human visual system usually contains three different types of cones used in perception of color. How are these cone types distributed across the retina of the eye to maximize color vision?
   A. short wavelength on the dorsal surface, long wavelength on the ventral surface
   B. long wavelength on the dorsal surface, short wavelength on the ventral surface
   C. short wavelength on the lateral region, long wavelength on the medial region
   D. long wavelength on the lateral region, short wavelength on the medial region
   E. more or less randomly

7. The primary function of amacrine cells within the retina is to:
   A. convert light energy into neural activity.
   B. process color vision.
   C. perceive shape.
   D. transmit information between ganglion and bipolar cells.
   E. All of the answers are correct.

8. Which of the following best describes the representation of visual information from our environment in the brain?
   A. Approximately half of the information from our visual field is represented in the brain.
   B. All information from the left eye is represented in the right brain.
   C. All information from the left eye is represented in the left brain.
   D. Approximately half of the information from the left eye is represented in the left brain.
   E. Visual information from our environment enters the left and right brain randomly.

9. Which of the following does *not* receive visual input via a primary visual pathway?
   A. frontal lobe
   B. occipital lobe
   C. parietal lobe
   D. temporal lobe
   E. All of the answers are correct.

10. Which of the following is *not* found in the occipital lobe of the human brain?
    A. primary visual cortex
    B. extrastriate visual cortex
    C. lateral geniculate nucleus
    D. blobs
    E. interblobs

11. A disproportionately large part of the __________ cortex is dedicated to processing information from the _________.
    A. frontal; periphery of the retina
    B. frontal; fovea
    C. occipital; periphery of the retina
    D. occipital; fovea
    E. temporal; periphery of the retina
    F. temporal; fovea

12. At what level of the visual system are shapes perceived?
    A. rods
    B. cones
    C. ganglion cells
    D. thalamus
    E. cortex

13. Utilization of luminance contrast by the visual system would be best employed under what condition?
    A. trying to view a moving object
    B. trying to establish the location of the edge of an object
    C. trying to distinguish a difference in the shade of two colors
    D. trying to determine which of two objects was larger
    E. trying to establish which of two objects was closer

14. Which of the following characteristics differentiates hypercomplex cells from complex cells?
    A. Hypercomplex cells are sensitive to color.
    B. Hypercomplex cells are sensitive to a moving bar of light.
    C. Hypercomplex cells are located in the visual cortex.
    D. Hypercomplex cells have a strong inhibitory area at one end of their receptive field.
    E. Hypercomplex cells are orientation detectors.

15. Using the trichromatic theory of color vision, how would you explain the most common type of color blindness experienced by humans?
    A. It is caused by absence of retinal receptors.
    B. It is caused by an absence of all cones.
    C. It is caused by an absence of one type of cone.
    D. It is caused by an absence of blobs.
    E. The basis for this color blindness can only be explained using opponent-process theory.

16. Which of the following cortical cells or regions are thought to process color vision?
    A. simple cells
    B. complex cells
    C. hypercomplex cells
    D. blobs
    E. interblobs

17. Scotomas, or blind spots in the visual field, may be caused by which of the following?
    A. loss of rods
    B. loss of cones
    C. lesions of the occipital lobe
    D. lesions of the temporal lobe
    E. All of the answers are correct.

18. Visual-form agnosia may be caused by extensive lesions of the lateral occipital region. Which of the following is the correct definition for the general term *agnosia*?
    A. not knowing
    B. not seeing
    C. no color
    D. no form
    E. no eyes

19. Which of the following is considered to be among the most common permanent neurological symptoms of carbon-monoxide poisoning?
    A. migraine headache
    B. visual agnosia
    C. optic ataxia
    D. color blindness
    E. blindness

20. Which of the following would you expect to observe in a person who suffers from an optic ataxia?
    A. inability to recognize objects
    B. unimpaired ability to handle objects
    C. damage to the retinal surface
    D. All of the answers are correct.
    E. None of the answers is correct.

Short-Answer Questions

Answer each of the following questions with a brief but complete written answer based on information from your text.

1. Neurons are activated by chemical neurotransmitters. Sensory receptors are specialized neurons that are activated by environmental energy. What is the source of this energy for the following sensory systems: vision, auditory, taste, olfaction?

2. Beyond our range of color perception are waves of electromagnetic energy that, though invisible to our sensory system, may be perceived by other animals. What are the terms used for wavelengths too short and too long for us to perceive?

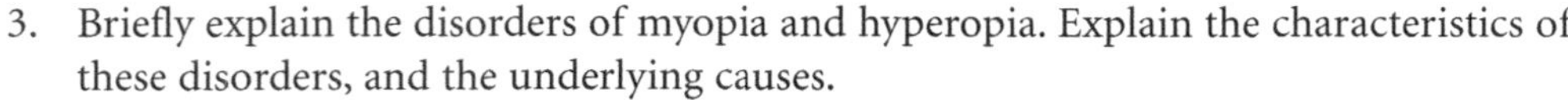

3. Briefly explain the disorders of myopia and hyperopia. Explain the characteristics of these disorders, and the underlying causes.

4. Most individuals have three distinct types of cones. Briefly explain what feature of these cones differs and why this difference is important for our visual perception.

5. Jerison's principle of proper mass states that the amount of neural tissue responsible for a particular function is directly proportional to the complexity of that function. With this in mind, explain how neural tissue for visual processing is distributed in relation to cells in the retina.

6. You have learned that the corpus callosum is an important structure for transmitting information between the two hemispheres. There are, however, very few corpus callosum connections between the occipital lobes. Why are there so few connections, and what is the most likely explanation for the few connections that are made between these visual regions?

7. Briefly describe the difference between simple cells, complex cells, and hypercomplex cells of the primary visual cortex.

8. Briefly describe the physiological function of cones proposed in the trichromatic theory of color vision.

9. People who are color blind generally do see some color, they simply have difficulty distinguishing between some colors. What is the physiological basis for color blindness?

10. Visual-form agnosia can result in an inability to recognize objects, but not to interact with them. Optic ataxia, on the other hand, does not impair ability to recognize objects, but can severely impair ability to reach for and manipulate objects. Briefly explain the importance of characterizing these two disorders in terms of understanding visual processing.

Matching Questions

Complete each of the following matching questions based on information from your text.

1. Match the following structures of the eye with the appropriate feature or function.

A. Cornea	___ Contains photoreceptors
B. Lens	___ Directs image onto the fovea
C. Iris	___ Controls amount of light entering eye
D. Retina	___ Contains blood vessels entering eye
E. Blind spot	___ Outer covering of eye

2. Indicate whether the following is a characteristic of rods (R) or cones (C).

___ used primarily for night vision
___ used for color vision
___ highest density found in the fovea
___ used for acute vision
___ long and slender in shape
___ most of the receptors in the eye are this type

3. Indicate whether the following is a characteristic of magnocellular cells (M) or parvocellular cells (P).

___ receive input primarily from rods
___ found in the periphery of the retina
___ are sensitive to color
___ are the smaller of the two ganglion cells
___ are more sensitive to light

4. Match each of the following disorders of vision with the appropriate feature or symptom.

A. Scotomas	___ Deficit in reaching using visual guidance
B. Homonymous hemianopia	___ Usually compensated for by nystagmus
C. Color blindness	___ Caused by complete lesion of one optic tract
D. Visual-form agnosia	___ More common in males than females
E. Optic ataxia	___ Inability to recognize objects

5. Indicate from first (1) to last (5) the pathway of neural signals through the visual system.

___ Optic chiasm
___ Occipital lobe
___ Lateral geniculate nucleus
___ Fovea
___ Temporal lobe

Diagrams

1. In the diagram of an eye below label the following structures: cornea, lens, iris, sclera, retina, fovea, blind spot, pupil, optic nerve.

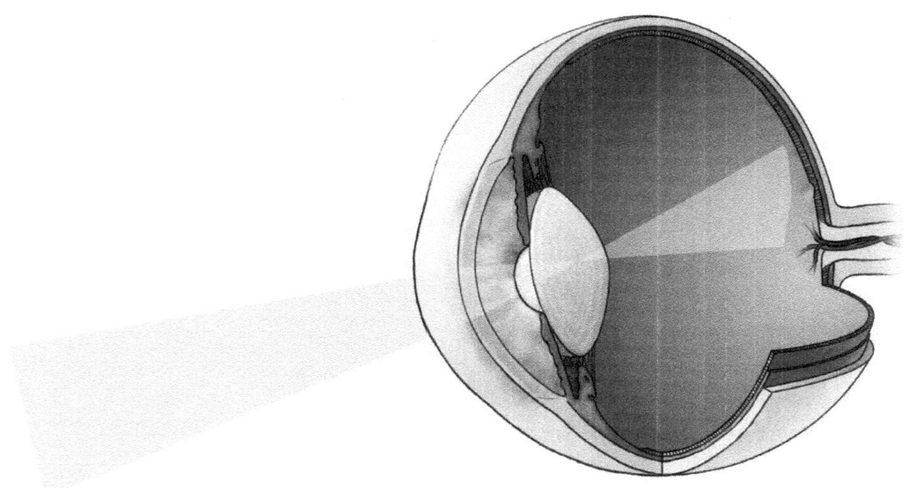

2. In the diagram of a retina below, label the following cells: cone, rod, ganglion cell, horizontal cell, amacrine cell, bipolar cell.

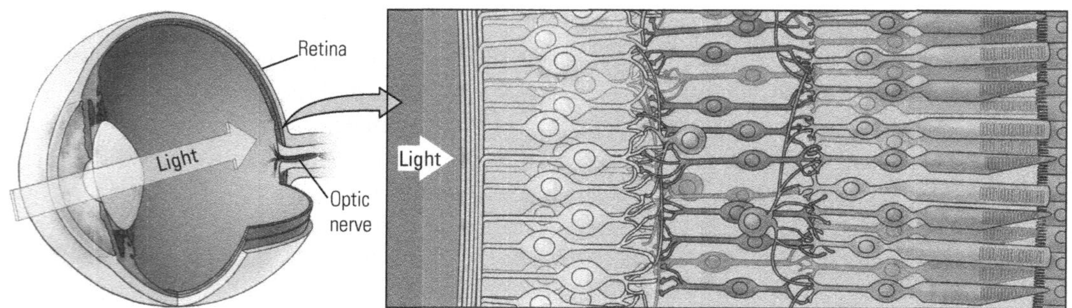

3. Show the pathway for visual information as it moves from the visual field to the brain. Use particular care when showing the pathway's movement through the optic chiasm.

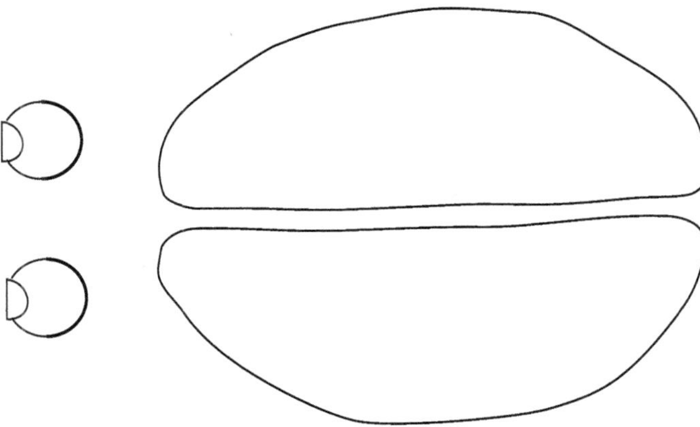

4. Based on the diagram below showing cone responses in the color spectrum, what color would you expect to see at wavelengths indicated by A and B?

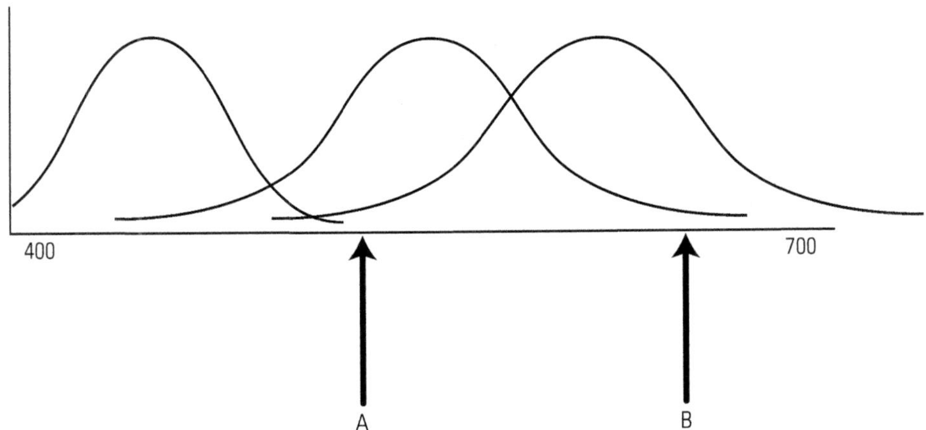

5. Regarding the columnar organization of cells in the primary visual cortex, the diagram below shows a vertical bar of light to which cells in one column respond. Indicate the approximate stimuli required to evoke responding from the adjacent columns.

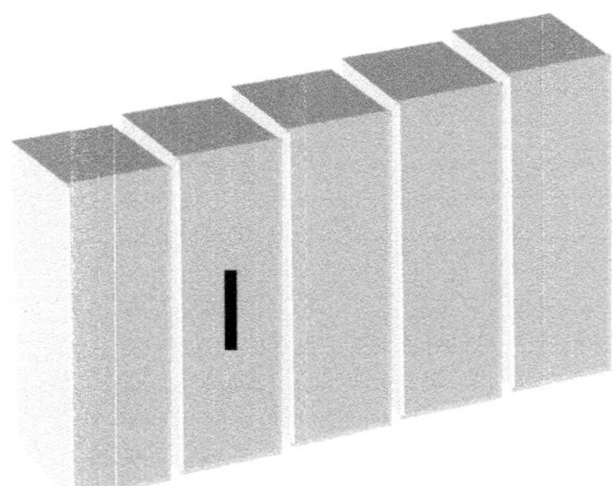

The Web

Consider using the following Web sites for additional information on some of the topics from this chapter:

1. Blindness resource site: www.nyise.org/blind.htm
2. Check to see if you're color blind: colorvisiontesting.com/
3. Vision Sciences Society: www.visionsciences.org/
4. Eye and visual system anatomy: tedmontgomery.com/the_eye/
5. Visual Illusion Gallery: http://dragon.uml.edu/psych/illusion.html

CROSSWORD PUZZLE

Across

1. Clear protective covering over eye
4. Blind _____, where axons leave the eye
6. With 22 across, this is the other
8. Prefix with chiasm, nerve, and trait
10. Dark staining cortex regions thought to process color
12. Visual part of the thalamus; abbr.
13. The "L" in LGN
14. M cells are magnocellular; P cells are _____
16. Light-gathering organs
17. Areas near 10 across that don't stain dark
22. Opponent-process is an example of one of them for color vision
25. They are found among 26 across and horizontal cells
26. Cells that receive input from 27 down
28. The "G" in LGN
31. Sad eye excretion
32. _____ colliculus receives visual input
33. Where the visual receptors are located in your eyes
34. Smell with your nose, see with your _____
35. Outer layer of occipital and temporal lobes that process vision
36. Like 1 down, but with a strong inhibitory area at one end of the field

Down

1. Visual cortex cells that respond to a moving bar of light
2. Term for an inability to visually guide movements
3. The "higher" stream for visual information
5. Straite cortex is sometimes referred to as the _____ visual cortex
7. Interpreted from wavelengths of light
9. Visual system is to see, as auditory system is to _____
11. It directs light to the fovea
15. Secondary visual cortex is sometimes referred to as _____ cortex
18. Without sight
19. Number of eyes on a cyclops
20. Means "stripe," used to describe visual cortex sometimes
21. Holes in the eyes through which light passes
23. Cortical cell that responds best to a stationary bar of light
24. 24 down; singular
27. Most abundant visual receptor
29. May be blue, green, or brown . . . not a cone!
30. Another term for 4 across, optic _____
33. On/off pairing for opponent process; abbr.

CHAPTER

10 How Do We Hear, Speak, and Make Music?

CHAPTER SUMMARY

As with vision, the perception of sound requires a transformation of environmental energy into neural energy. However, unlike the visual system, which utilizes electromagnetic energy, the auditory system utilizes pressure waves formed from vibrating air molecules. Sound waves have three properties that we are able to discern: *frequency*, *amplitude*, and *complexity*. Frequency of sound waves is measured in *hertz* (cycles per second). Variations in hertz are perceived as differences in pitch. As with the visual system, the human range of audition is not infinite, and other animals use energy beyond the range humans can perceive. Humans hear from about 20 Hz (low pitch) to about 20,000 Hz (high pitch). Very few sounds are of a pure or single frequency. Rather, sounds are made up of combinations of frequencies called complex tones. Complexity is perceived as quality of tone patterns. Random sounds that are not interpreted as single frequencies or complex tone patterns are perceived as noise. Amplitude of sound waves is measured in *decibels*, and perceived as volume or intensity of sound by humans. Our auditory system is also designed to localize the source of sound production.

Perception of different sounds is a function of interpreting different sequences of neural activity. This interpretation is largely a process through which meaning about our environment is gained. The process of auditory analysis is far more complex than simply detecting the presence of sound. For example, we can understand speech at a rate of nearly 30 segments per second, which is approximately 5 times the rate at which we can interpret random sounds. We are also capable of discerning three perceptual qualities from music, including *timbre*, which allows us to differentiate numerous sources of a musical tone of the same pitch and volume.

Perception of sound begins in the ear where the *pinna* acts to catch and deflect sound waves into the external ear canal. Next the energy waves vibrate the *eardrum*, which in turn vibrates a series of small bones commonly called the *hammer*, *anvil*, and *stirrup* in synchrony with the sound waves. These bones then flex a small membrane called the *oval window* on the *cochlea* of the inner ear. Movement of the oval window results in movement of the fluid inside the cochlea. In the center of the cochlea, the *basilar membrane* contains approximately 15,000 *hair cells* that are the receptors for transforming physical energy (waves of fluid inside the cochlea) into neural energy.

Over 100 years ago Hermann von Helmholtz proposed that movement of different regions of the basilar membrane (and thus stimulation of different groups of hair cells) was perceived as different frequencies or pitches. In the 1960s George von Békésy supported this general idea by showing that high-frequency sounds produce greater displacement of hair cells near the base of the basilar membrane while low-frequency sounds produce the greatest displacement near the apex. As a region along the membrane is maximally displaced by waves of fluid, hair cells are displaced to varying degrees. Maximum displacement in one direction results in a maximum influx of calcium (depolarization) and a strong excitatory signal. Hair cell displacement in the opposite direction closes calcium channels, resulting in hyperpolarization. Integration of multiple signals along the membrane ultimately results in a signal perceived as a particular pitch. Loudness of all tones is determined by the intensity of the signal. That is, how much displacement of cells has occurred and how much neurotransmitter is released as a result of calcium influx.

As with the visual system, the auditory system also utilizes bipolar and ganglion cells to propagate the signal from the receptor cells. Projections from ganglion cells enter the medulla at the *cochlear nucleus*. The signal is then sent to the nearby *superior olive* and the *trapezoid body*, and from these structures to the *inferior colliculus* and then the *medial geniculate nucleus*. Ultimately, auditory information is processed in cortical regions. Among these cortical regions is the *planum temporale*, which is physically larger in the left hemisphere than in the right hemisphere of right-handed individuals. The planum temporale includes *Wernicke's area*, a region that is known to be important in comprehension of speech.

Understanding how hair cells are arranged along the basilar membrane to respond maximally to a particular pitch (known as *tonotopic representation*) has allowed researchers to develop cochlear implants. These tiny devices contain a microphone to receive sound and a series of wires that are differentially stimulated based on the pitch of the sound received. Electrically stimulating auditory neurons ultimately produces a signal similar to stimulation encountered in an intact inner ear, making these devises relatively effective in restoring some aspects of auditory perception to many deaf individuals.

The one known exception to tonotopic representation of the basilar membrane is that very low tones (<200 Hz) do not maximally displace a region along the membrane, but rather, move a length of the apex of the membrane in synchrony with the waves. For these low tones, the system appears to simply utilize frequency of movement of the membrane apex to determine differences in pitch.

Location of sounds is determined using two methods. First, the timing of sounds stimulating receptors is determined. Receptors in the ear nearest to the sound source fire slightly ahead of those on the other side of the head. Second, high-frequency sounds are noticeably dampened by the head when entering the ear opposite the source of the sound. In both cases, sounds coming from directly in front, behind, above, or below the head are particularly difficult to localize. When this happens, there is a natural tendency to turn or tilt the head slightly to help localize the sound source.

Regarding interpretation of complex sounds, there appear to be regions in the right and left temporal cortex for assessing music and language, respectively. Damage to these regions can result in specific disruption of ability to interpret these complex stimuli. Also interesting is the finding that damage to what is considered the "language region" in monkeys results in a disruption of ability to recognize vocalizations of that species.

In the 1960s Noam Chomsky proposed that language production and interpretation had features of an inherited trait, suggesting that humans were predisposed to learning language. Since that time additional evidence has supported this contention. For example, elegantly complex languages are learned with very little apparent effort by children of all cultures, and language structure develops with no formal training. Children also show a sensitive period for learning language, after which time learning language is very difficult.

Language appears to be centered in at least two regions within the human cortex (usually in the left hemisphere). *Broca's area*, located at the posterior region of the frontal lobe, is responsible for language production. Damage to this area results in a disruption of normal speech with few obvious effects on language comprehension and is known as *Broca's aphasia*. Wernicke's area, located in the posterior region of the temporal lobe, is responsible for language comprehension. Damage to this area results in disruption of language comprehension. Speech production is fluent, but usually nonsensical. This type of language disruption is known as *Wernicke's aphasia*. In the 1930s Wilder Penfield expanded the understanding of speech centers using electrical stimulation of the human brain in awake subjects. Penfield was able to induce vocalizations in subjects by stimulating regions he termed *supplementary speech areas*. He was also able to induce aphasias and speech arrest by stimulation of four different cortical regions (including Broca's and Wernicke's).

With technological advances, including the refinement of *positron emission tomography (PET)*, researchers are now able to view brain activity during speech in a noninvasive manner. Such studies, comparing brain activity during speech to brain activity in a control state, have shown separate regions for analysis of simple and complex auditory stimulation. While simple stimulation affects primarily A1 regions, complex stimuli (e.g., syllables) produce the greatest activity in secondary auditory regions. Although all auditory stimuli are processed in both hemispheres, the left hemisphere generally shows a greater response. PET scans have also shown that Broca's area becomes active not only when producing speech, but also during speech comprehension.

Analysis of music is generally thought of as a function of the right hemisphere. However, PET studies have shown that some aspects of music comprehension appear to require left hemisphere input. The distinction may be that the left hemisphere is involved with more sophisticated analysis of music (i.e., discerning sequences of pitches).

Auditory communication is prevalent in many species. Birds provide a particularly useful model for study, in part because birdsong parallels human language on several levels. There is a diversity and complexity of song across many species. The final song utilized by a bird is affected by experiences during development, particularly during a sensitive period early in life. Control of birdsong appears to be asymmetrical, controlled primarily by the left hemisphere. Finally, there are separate structures for producing and interpreting song in the bird brain.

Bats also utilize auditory signals, but primarily for the purpose of echolocation rather than for communication. Bats emit a range of high-frequency noises from which echoes can be analyzed to guide flight. Interestingly, the bats' auditory system has developed a "cochlear fovea," or a region that is particularly responsive to the range of sounds used for echolocation. The brain region dedicated to analysis of this range is particularly large, aiding in complex analysis of echoes for location, movement, and even surface texture.

KEY TERMS

The following is a list of important terms introduced in Chapter 10. Give the definition of each term in the space provided.

Sound

Frequency

Hertz

Perfect pitch

Amplitude

Decibels

Pure tones

Complex tones

Noise

Pitch

Prosody

Quality

Timbre

Anatomy of the Auditory System

Pinna

External ear canal

Ossicles

Hammer

Anvil

Stirrup

Eardrum

Oval window

Cochlea

Basilar membrane

Hair cells

Tectorial membrane

Auditory Receptors

Hair cell

Bipolar cells

Cochlear nucleus

Olivary complex

Trapezoid body

Inferior colliculus

Medial geniculate nucleus

Auditory Cortex

Heschl's gyrus

Planum temporale

Wernicke's area

Lateralization

Insula

Neuronal Activity and Hearing

Tonotopic representation

Tuning curve

Cochlear implants

Anatomy of Language and Music

Creolization

Broca's area

Broca's aphasia

Wernicke's aphasia

Aphasia

Supplementary speech area

Positron emission tomography (PET)

Echolocation

KEY NAMES

The following is a list of important names introduced in Chapter 10. Explain the importance of each person in the space provided.

Hermann von Helmholtz

George von Békésy

Noam Chomsky

Paul Broca

Karl Wernicke

Wilder Penfield

PRACTICE TEST

Multiple-Choice Questions

Answer each of the following multiple-choice questions with the best possible answer based on information from your text.

1. What was found in the cave of Neanderthals that changed the way anthropologists and neuroscientists viewed language and music development in this early culture?
   A. paintings on the cave wall
   B. pottery with elaborate inscriptions
   C. a primitive flute
   D. a primitive drum
   E. a primitive writing utensil

2. What form of energy do animals perceive as sound?
   A. waves of air molecules
   B. waves of electromagnetic energy
   C. waves of volatile chemicals
   D. waves of heat energy
   E. All of the answers are correct.

3. The frequency range in which humans can distinguish sound is approximately ________ hertz.
   A. 2–200
   B. 20–200
   C. 20–2,000
   D. 20–20,000
   E. 200–20,000

4. A person with perfect pitch is able to do which of the following?
   A. perceive a larger range of frequencies than normal individuals
   B. perceive sounds at lower amplitudes than normal individuals
   C. identify by sound any note on the musical scale
   D. sing extraordinarily well
   E. play musical instruments extraordinarily well

5. A sound characterized as 80 decibels and 80 hertz could be described as which of the following?
   A. a high-pitched soft sound
   B. a high-pitched loud sound
   C. a low-pitched soft sound
   D. a low-pitched loud sound
   E. You cannot describe pitch and volume based on these measures.

6. Which of the following is *not* true of speech perception?
   A. It is affected by experience.
   B. It is mediated in large part by the temporal lobe.
   C. It can occur at rates faster than perception of nonspeech sounds.
   D. It shares many features of music perception.
   E. It is most fully developed in cultures that produce rapid speech.

7. Which of the following structures is *not* located inside the cochlea?
   A. basilar membrane
   B. tectoral membrane
   C. inner hair cells
   D. outer hair cells
   E. cochlear nucleus

8. Which of the following best describes the way in which high-pitch sounds are transformed into neural signals along the basilar membrane?
   A. They maximally displace hair cells near the apex of the basilar membrane.
   B. They maximally displace hair cells near the base of the basilar membrane.
   C. They maximally displace hair cells near the middle of the basilar membrane.
   D. They displace hair cells equally along the entire length of the basilar membrane.
   E. They do not displace any hair cells along the basilar membrane.

9. Displacement of hair cells along the basilar membrane initiates an excitatory neural signal by which of the following?
   A. causing an influx of calcium
   B. causing an influx of sodium
   C. causing an influx of potassium
   D. causing an efflux of chloride
   E. causing an efflux of sodium

10. At what region do axons from the auditory bipolar cells enter the brain?
    A. temporal lobe
    B. frontal lobe
    C. parietal lobe
    D. occipital lobe
    E. brain stem

11. While Wernicke's area is specialized for language comprehension in one hemisphere, the same area in the other hemisphere is specialized for which of the following functions?
    A. language comprehension
    B. music comprehension
    C. speech production
    D. producing written language
    E. reading

12. Tonotopic representation predicts that:
    A. humans can identify any rhythmic sound.
    B. perfect pitch is innate.
    C. each hair cell is maximally responsive to a particular frequency.
    D. audition is difficult to achieve without visual representation of the object.
    E. All of the answers are correct.

13. Which of the following mechanisms of neural signaling is used to indicate an increase in the volume of a sound?
    A. an increase in the size of action potentials
    B. a decrease in the size of action potentials
    C. an increase in the frequency of cell firing
    D. a decrease in the frequency of cell firing
    E. Perception of increased volume cannot be accounted for by neural firing.

14. What technique could you employ to help localize sounds that are difficult to pinpoint because they arrive at the same time with the same intensity at both of your ears?
    A. tilting your head
    B. holding perfectly still
    C. closing your eyes
    D. nodding your head up and down
    E. All of the answers are correct.

15. Which of the following is *not* true of language and language development?
    A. Language acquisition is nearly effortless for children.
    B. Technically advanced cultures tend to have the most complex languages.
    C. If language is not learned in the first 6 years of life, skills will be severely compromised.
    D. Children do not learn the structure of language from their parents.
    E. All human populations have language.

16. Which of the following is true of Broca's area?
    A. It is primarily responsible for language production.
    B. It is located in the frontal lobe.
    C. It is generally localized to the left hemisphere.
    D. PET studies show that it becomes active during language/sound discrimination tasks.
    E. All of the answers are correct.

17. Which of the following is true of the electrical brain stimulation studies conducted by Wilder Penfield?
    A. He was able to evoke complex speech patterns from patients.
    B. He was able to induce complex auditory signals perceived by patients.
    C. He was able to induce both aphasia and complete arrest of speech.
    D. He identified approximately 30 cortical areas for language comprehension and production.
    E. All of the answers are true.

18. Which of the following is true of neural analysis of music?
    A. Most analysis of music occurs in the right hemisphere.
    B. Most analysis of music occurs in the left hemisphere.
    C. Analysis of music occurs approximately equally in the left and right hemispheres.
    D. Broca's area is critically involved in analysis of music.
    E. Wernicke's area is critically involved in analysis of music.

19. Which of the following is *not* a feature of birdsong that parallels a feature of human language development?
    A. There exists a sensitive period for learning.
    B. There are structures for both production and comprehension.
    C. There are regional dialects for the same song/language.
    D. The left hemisphere appears responsible for the majority of neural processing.
    E. Neural structures responsible for song/language are sexually dimorphic and larger in males.

20. Which of the following features can a bat detect using echolocation?
    A. location of an object
    B. velocity of a moving object
    C. surface texture of an object
    D. distance of an object
    E. All of the features can be detected.

Short-Answer Questions

Answer each of the following questions with a brief but complete written answer based on information from your text.

1. Both complex tones and noise are made up of combinations of frequencies. Briefly describe the difference between complex tones and noise as they are perceived by humans.

2. Describe in general terms the theory of pitch perception as proposed orginally by Hermann von Helmholtz and later modified by George von Békésy.

3. Cochlear implants have a small microphone that picks up sounds in the environment. Briefly describe how these implants transfer that information to the brain.

4. Briefly describe the two ways in which our auditory system may localize the source of sound.

5. Noam Chomsky shook up the world of language research with his theory that language may in fact be an inherited trait. Give at least two examples of characteristics of language that support Chomsky's contention.

6. Early slave traders attempted creolization with a language they created and called "pidgin." Why did pidgin fail as a language, and what does its failure tell us (in general) about the development of language?

7. Briefly describe deficits you might observe in an individual who has damage to Broca's area.

8. Briefly describe deficits you might observe in an individual who has damage to Wernicke's area.

9. List at least three factors that can affect the development of birdsong in young birds.

10. Describe the concept of a "cochlear fovea" as it relates to the use of echolocation in bats.

Matching Questions

Complete each of the following matching questions based on information from your text.

1. Match the following terms with the appropriate feature or description.

A. Frequency	___ Measured in decibels
B. Amplitude	___ Primarily right hemisphere
C. Location	___ Measured in hertz
D. Language	___ Time difference between ears
E. Music	___ Posterior temporal lobe

2. Indicate from first (1) to last (5) structures that are utilized along the auditory pathway.

___ Ganglion cells
___ Inferior colliculus
___ Bipolar cells
___ Cochlear nucleus
___ Superior olive

3. Match the following sound descriptions to their appropriate waveforms: A) Low pitch loud, B) Low pitch quiet, C) High pitch loud, D) High pitch quiet

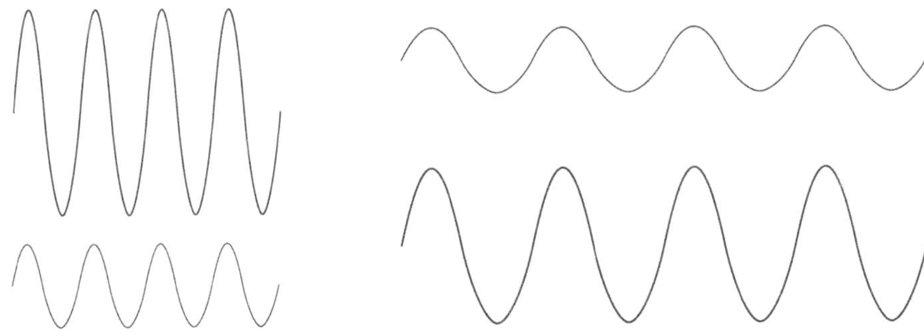

4. Match the following structures associated with hearing with the brain structures in which they are located.

A. Wernicke's area	___ Cortex
B. Medial geniculate nucleus	___ Hindbrain
C. Inferior colliculus	___ Cochlea
D. Olivary complex	___ Thalamus
E. Round window	___ Midbrain

5. Match the following location from which a sound originates with the appropriate perception strategy.

A. Looking straight ahead, quieter in right ear	___ From the right
B. Looking left, quieter in left ear	___ From ahead
C. Looking right, arriving first at the right ear	___ From behind
D. Looking behind, arriving first in left ear	___ From the left

Diagrams

1. On the diagram below, identify the following inner ear structures: Inner hair cells, Outer hair cells, Basilar membrane, and Tectorial membrane.

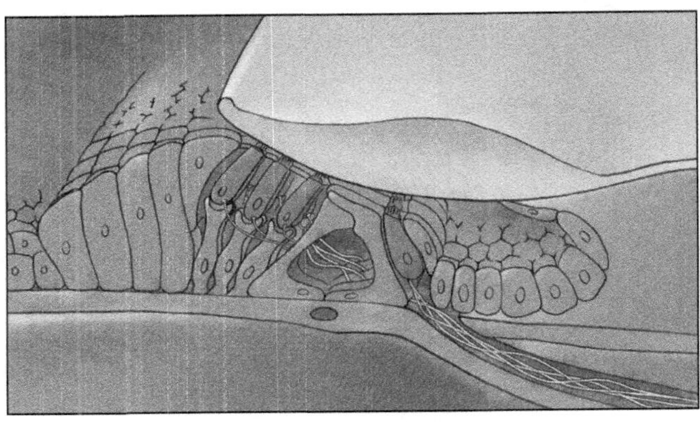

2. Show where you expect maximum displacement of hair cells along the basilar membrane below for the following instruments: Tuba, Flute, Trumpet

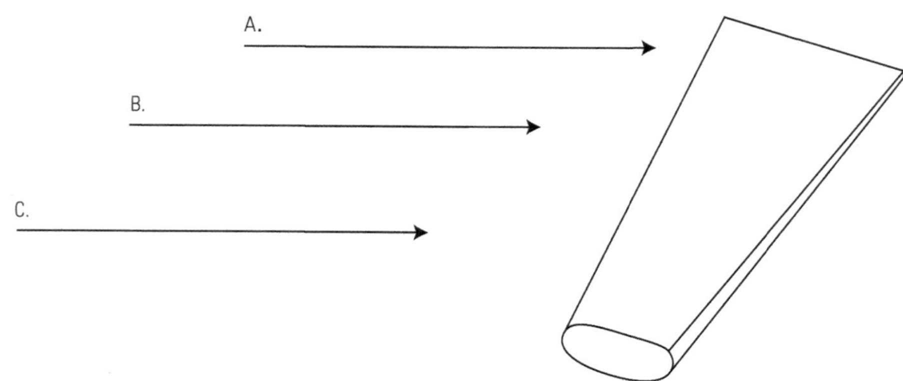

3. Which of the following sound waves would be the most difficult for this person to localize?

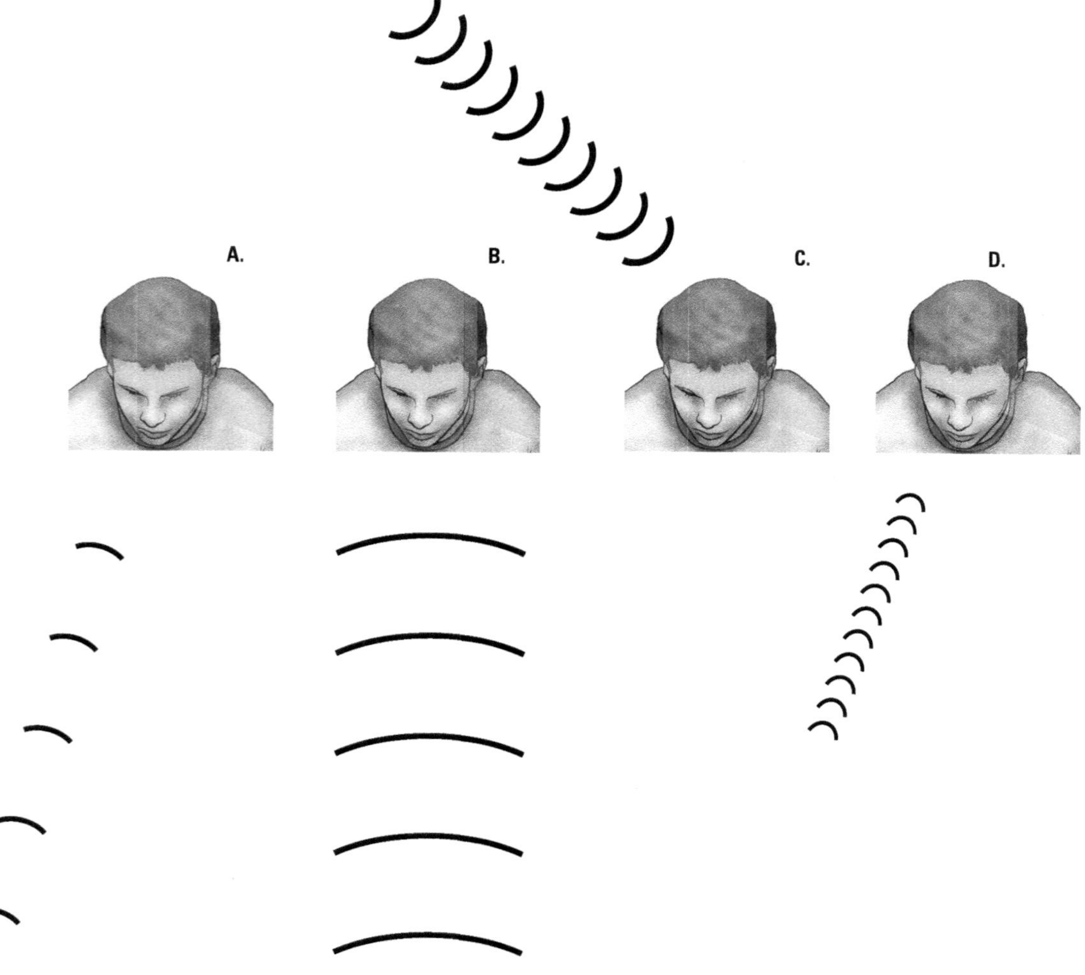

4. On the figure below, identify the approximate location of the following regions: Motor area for face, Broca's area, Wernicke's area, Arcuate fasciculus.

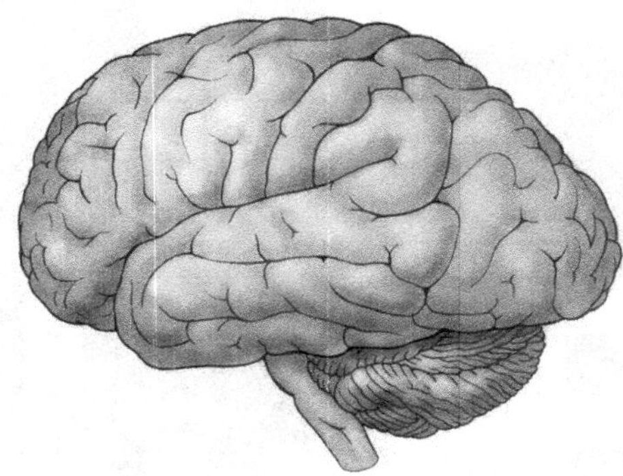

5. Arrange the following sounds in their approximate location along the dB scale provided below: A) Your professor's voice when lecturing, B) A fire alarm in your building, C) A person whispering at the desk next to you, D) A low-flying helicopter passing over your building

---

0 20 40 60 80 100 120 140 160 180 200

The Web

Consider using the following Web sites for additional information on some of the topics from this chapter:

1. National Association of the Deaf: www.nad.org
2. American Academy of Audiology: www.audiology.org/
3. Graphic tour of the inner ear: www.vimm.it/cochlea/index.htm
4. National Aphasia Association: www.aphasia.org/
5. Information on cochlear implants: www.nidcd.nih.gov/health/hearing/coch.asp

CROSSWORD PUZZLE

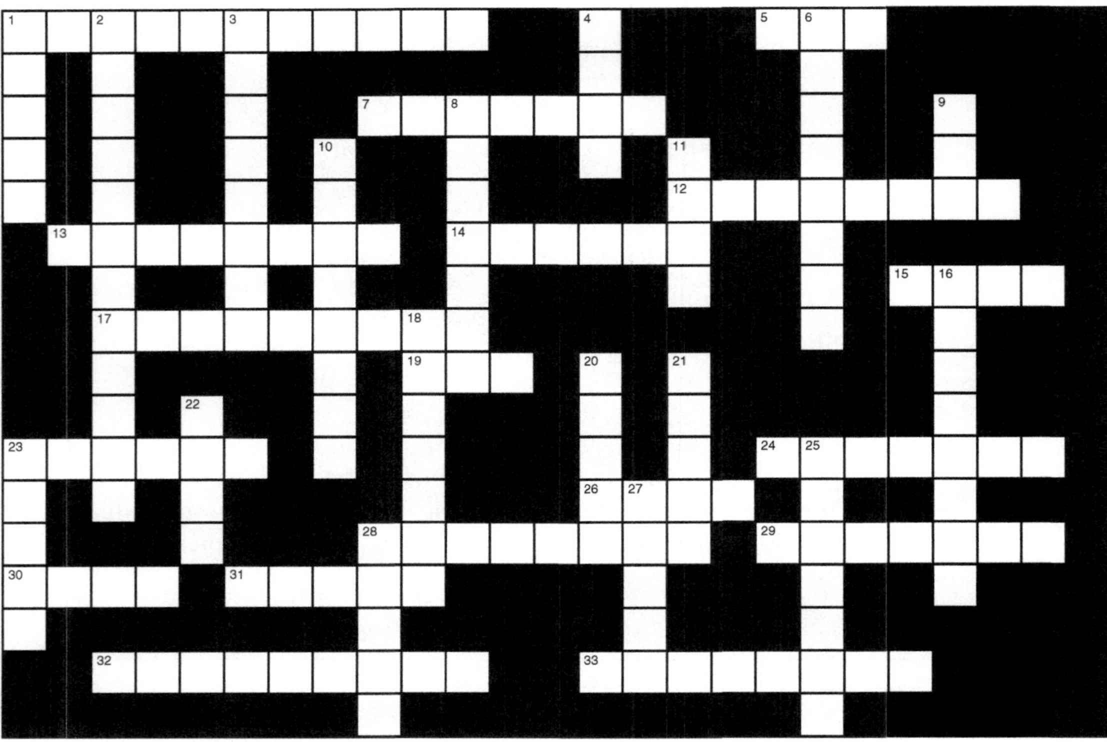

Across

1. Waves used to detect pitch
5. Loss of hearing may come with an increase in this
7. Vision utilizes the lateral geniculate _____; hearing utilizes the medial geniculate _____
12. Cochlear _____ may help restore hearing
13. Another term for 17 across
14. One of the 3 small ear bones
15. Without hearing
17. Another term for volume
19. Organ designed to capture and direct sound waves
23. Region essential for language production
24. Term for disrupted speech or language comprehension
26. Canaries use it instead of language
28. He produced 24 across, electrically stimulating brains of conscious patients
29. Rhythm of speech
30. It doesn't sing, but it does caw
31. One of the 3 small ear bones
32. Basilar and tectoral, e.g.
33. Region essential for language comprehension

Down

1. Tuning _____ are sometimes used to produce 1 across
2. Bats use it to guide flight
3. _____ ear canal leads to 4 down
4. Common name for tympanic membrane is the ear _____
6. Vision and hearing both use these cells right after bipolar cells
8. Snail-shaped inner ear compartment
9. Scan for watching brain activity; abbr.
10. The first word abbreviated in 9 down
11. 30 across is an example of one
16. The second word abbreviated in 9 down
18. Measure of loudness
20. Random strings of 21 down
21. What you perceive from compressed air waves
22. Type of cell found in the inner ear
23. The researcher who identified and named a region essential for speech
25. _____ pitch allows you to hear and then identify any note on the scale
27. Inferior _____ is essential for processing auditory input
28. That outer fleshy part of 19 across

CHAPTER

11 How Does the Nervous System Respond to Stimulation and Produce Movement?

CHAPTER SUMMARY

Movement involves many brain regions arranged in a *hierarchical organization*. This means that movement controlled by higher brain regions are more complex and under greater voluntary control than those controlled by lower regions. John Hughlings-Jackson proposed three levels of hierarchical organization, mediated roughly by the forebrain (highest level), the brainstem, and the spinal cord (lowest level). He further suggested damage to the higher levels could produce dissolution (the opposite of evolution), whereby primary movements would become very simple under the control of lower levels. In the 1950s Karl Lashley suggested that the central nervous system was capable of producing *motor sequences* that could control streams of rapid complex behaviors (such as those required for playing a musical instrument or producing speech) without sensory feedback. It was later determined that the frontal lobe indeed has four primary motor regions, including one for planning complex behaviors (prefrontal cortex) and two for producing appropriate movement sequences (supplementary and premotor cortex). The fourth region, primary motor cortex, is involved in executing the details of movements as planned and arranged by the other three regions.

The brainstem is responsible for species-typical behavior. This is evidenced when regions of brainstem are stimulated with electrodes in awake animals. Depending on the specific region stimulated the animal may respond with aggressive behaviors, fear behaviors, feeding behavior and so on. Damage to the brainstem during fetal development or in early infancy can result in a condition known as *cerebral palsy* in which a person of normal intellect may exhibit profound movement disruption including, in some cases, an inability to speak.

The primary function of the spinal cord in the execution of movement is relaying information from the brain to the appropriate skeletal muscles. Damage to the spinal cord, depending on the location, can result in *paraplegia* or *quadraplegia*. In addition to this function, the spinal cord also contains some simple reflexive motor patterns, including a walking pattern for the legs, and a *scratch reflex*, which can be seen in many animals.

Electrical stimulation studies in animals and humans have led to a mapping of the motor cortex relative to the body regions it controls. The distribution is topographically similar to the body, with control of head adjacent to control of neck, adjacent to control of shoulder, and so on. However, the amount of tissue devoted to movement of body regions is highly disproportionate. Small body regions that are utilized for highly skilled and rapid movements (e.g., lips, hands) are controlled by the greatest amount of neural tissue. A *homunculus* is often used to show the disproportionate distribution of tissue as it relates to specific body regions. In general, damage to a portion of the motor cortex associated with a particular region of the body will result in motor deficits in that region. Early researchers, including Wilder Penfield, found short duration electrical stimulation of specific motor cortex regions produced simple movement of the associated body part. More recently, researchers have discovered that longer duration stimulation of some regions can produce more complex motor patterns, including reaching and posturing behaviors.

Neural recordings show that motor cortex regions are utilized prior to and during movement patterns. If cortical regions controlling movement are damaged, there may be some reorganization, with nearby regions assuming control of muscles normally controlled by the damaged area. Furthermore, this reorganization is enhanced if the body region affected by the neural damage is forced into use during the recovery period, a technique known as *restraint-induced therapy.*

The *corticospinal tracts* (or *pyramidal tracts* as they are sometimes called) are the main pathways from the motor cortex to the brainstem and spinal cord. The somas of motor neurons are located in the ventral horn of the spinal cord, with axons that stretch to (and innervate) muscles throughout the body. Motor neurons innervating the arms and legs are controlled by the motor cortex on the opposite side of the brain (contralateral). Motor neurons that innervate the trunk regions are controlled by the same side of the brain (ipsilateral).

The *basal ganglia* are several nuclei of the forebrain that make connections with the overlying motor cortex and with the midbrain. The *caudate* comprises the largest portion of this region and receives primary input from the cortex and dopaminergic input from the substantia nigra. Damage to the basal ganglia results in a full spectrum of motor disruption. Loss of dopamine input from the substantia nigra results in a *hypokinetic* state, the most common example being the rigidity associated with Parkinson's disease. Direct damage to the caudate results in a *hyperkinetic* state, as seen in the uncontrollable limb movements associated with Huntington's chorea or the uncontrollable tics and vocalizations associated with Tourette's syndrome. The "volume theory" of basal ganglia control over behavior suggests that these structures mediate the extent (volume) of movement planned and executed by cortical regions.

The *cerebellum* contains nearly half of the neurons in the central nervous system and is critical for acquiring and maintaining skilled motor patterns. It also controls many aspects of balance and eye movement. The primary function of the cerebellum is timing movements and maintaining movement accuracy. Repetition of movement patterns results in improved motor skills with enhanced cerebellar control of the motor pattern.

The motor system is intimately integrated with the *somatosensory system.* Sensory feedback from muscles sends signals to the primary *sensory neocortex* located directly adjacent to the motor cortex. The vestibular system, located within the middle ear, provides primary information to the cerebellum about balance. Sensory receptors are located in *hairy skin* and *glabrous* (hairless) *skin.* Glabrous skin includes the lips and hands and is more sensitive to stimuli than hairy skin that covers the majority of our bodies. Dendrites of sensory receptor cells may be stimulated by pressure for the sense of *hapsis* (touch), by chemicals for the sense of *nocioception* (pain), or by stretch for the sense of *proprioception* (body location). Cell bodies for sensory receptors are located in the *dorsal root ganglion* of the spinal cord, with long dendrites reaching the target sites of the body. Proprioceptive and haptic information is conveyed rapidly along large myelinated axons, whereas nocioceptive information moves much slower along thin axons with less myelin.

In rare cases individuals have become deafferented (losing all sensory information from the body). When this happens, movement is profoundly affected and many motor patterns (including limb movement) require visual guidance. In this regard, sensory input can be thought of as the "eyes" that guide movement.

The *dorsal spinothalamic tract* is made up of haptic-proprioceptive axons traveling from the brainstem through the *medial lemniscus* to the *ventrolateral thalamus*. From there, information is sent primarily to the somatosensory cortex, but also to the motor cortex. Nocioceptive axons comprise the *ventral spinothalamic tract* that follows a path similar to that of the dorsal tract.

The spinal cord contains some basic motor programs like the knee-jerk response evoked by a monosynaptic reflex. Within the spinal cord, the dorsal and ventral tracts may interact as well. The gate theory of pain suggests that signals from the ventral tract (pain) may be blunted by vigorous stimulation of the dorsal tract. This *pain gate* is noticed when we rub the area of a minor injury to reduce the sensation of pain. Opiates such as morphine are also thought to work by gating pain signals along this tract. In some instances, the sensation of pain in one region of the body is actually evoked by neural signals from a nearby region that does not contain pain receptors. This *referred pain* is obvious in heart attack victims who feel pain in shoulders or arms as a result of strong neural signals originating in the heart where there are no pain receptors.

The *vestibular system* provides sensory information about body position from the *semicircular canals* and the *otolith organs* of the middle ear. *Endolymph*, the fluid within the semicircular canals, flows and splashes in response to body movements. Endolymph movement in turn bends hair cells that send neural signals about the direction and speed of body movement.

Somatosensory information is processed in the somatosensory cortex located just posterior from the central sulcus and the motor cortex in a region called the post-central gyrus. As with the motor cortex, a homunculus can be used to represent the relative areas of cortex assigned to each body region sending somatosensory information. Like the motor homunculus, the sensory homunculus shows disproportionately large regions dedicated to hands, fingers, and lips. Damage to the somatosensory cortex impairs movements of associated regions, often resulting in *apraxia*, and is likely due to lack of feedback necessary for guiding movement. Apraxia is a disorder in which the motor pattern needed for a movement is intact, but the ability to sequentially execute the movement is lost. As with the motor cortex, somatosensory cortex appears capable of reorganization in response to damage or in response to loss of input from sensory neurons.

KEY TERMS

The following is a list of important terms introduced in Chapter 11. Give the definition of each term in the space provided.

Hierarchical Control of Movement

Neuroprosthetics

Motor sequences

Prefrontal motor cortex

Premotor cortex

Primary motor cortex

Cerebral palsy

Spinal cord

Paraplegia

Quadraplegia

Scratch reflex

Organization of the Motor System

Homunculus

Topographic organization

Corticospinal tracts

Pyramidal tracts

Lateral corticospinal tracts

Ventral corticospinal tracts

Ventral horn

Basal Ganglia and Cerebellum

Basal ganglia

Caudate putamen

Tourette's syndrome

Hyperkinetic

Hypokinetic

Cerebellum

Somatosensory system

Hairy skin

Glabrous skin

Nociception

Hapsis

Proprioception

Rapidly adapting receptors

Slowly adapting receptors

Dorsal root ganglion neurons

Deafferented

Dorsal spinothalamic tract

Dorsal column nuclei

Medial lemniscus

Ventrolateral thalamus

Somatosensory cortex

Ventral spinothalamic tract

Monosynaptic reflex

Gate theory of pain

Pain gate

Periaqueductal gray matter (PAG)

Phantom limb pain

Referred pain

Vestibular system

Semicircular canals

Otolith organs

Utricle

Saccule

Endolymph

Otoconia

Apraxia

KEY NAMES

The following is a list of important names introduced in Chapter 11. Explain the importance of each person in the space provided.

Karl Lashley

Wilder Penfield

Walter Hess

PRACTICE TEST

Multiple-Choice Questions

Answer each of the following multiple-choice questions with the best possible answer based on information from your text.

1. Motor regions of the neocortex receive input from which of the following?
   A. visual regions of the cortex
   B. the basal ganglia
   C. sensory systems
   D. cerebellum
   E. All of the answers are correct.

2. According to the theory of hierarchical organization of the motor system, which of the following structures would likely be considered at the top of the hierarchy?
   A. spinal cord
   B. forebrain
   C. brainstem
   D. basal ganglia
   E. substantia nigra

3. In what lobe of the cerebral cortex can you find the primary motor cortex, supplementary motor cortex and premotor cortex?
   A. occipital
   B. frontal
   C. parietal
   D. temporal
   E. These regions span all four of the lobes.

4. Hess's brainstem stimulation studies suggested that this region was important in producing which of the following?
   A. sleep
   B. fear responses
   C. aggressive responses
   D. grooming behavior
   E. species-typical behaviors

5. Christopher Reeve suffered from a condition known as:
   A. cerebral palsy.
   B. paraplegia.
   C. quadraplegia.
   D. apraxia.
   E. homunculitis.

6. Abnormal development of which of the following structures has been associated with cerebral palsy?
   A. brainstem
   B. primary motor cortex
   C. primary somatosensory cortex
   D. basal ganglia
   E. spinal cord

7. What structure controls the scratch reflex in a dog?
   A. spinal cord
   B. cerebellum
   C. basal ganglia
   D. motor cortex
   E. The structure controlling this reflex is currently unknown.

8. Which of the following would be represented as disproportionately large in a human motor cortex homunculus?
   A. hands
   B. arms
   C. legs
   D. trunk
   E. None of the above would be disproportionately represented.

9. Recent studies in nonhuman primates conducted to confirm Penfield's original studies mapping the homunculus have revealed which of the following?
   A. that no such mapping pattern can be deduced using this technique
   B. that no areas of the body are disproportionately represented in the motor cortex
   C. that two distinct homunculi may be represented in the motor cortex
   D. that as many as ten homunculi may be represented in the motor cortex
   E. that the human homunculi representation is nearly identical to that of the rat

10. The corticospinal tracts from motor cortex to spinal cord give rise to large bumps on each side of the ventral surface of the brainstem. The shape of large bumps is the basis for which common term for the corticospinal tracts?
    A. pentagonal tracts
    B. spherical tracts
    C. pyramidal tracts
    D. mogul tracts
    E. rough tracts

11. Limb regions of the motor cortex send fibers along the __________ region of the corticospinal tract, whereas trunk regions send fibers along the __________ region.
    A. lateral; ventral
    B. dorsal; medial
    C. medial; anterior
    D. posterior; lateral
    E. anterior; dorsal

12. Nudo and colleagues found that reorganization of the cortical map was highly responsive to _________ during the three-month period following discrete cortical lesions.
    A. light therapy
    B. drug therapy
    C. electrical brain stimulation
    D. All of the answers are correct.
    E. None of the answers is correct.

13. Which of the following is a prominent structure within the basal ganglia?
    A. cerebellum
    B. caudate putamen
    C. amygdala
    D. substantia nigra
    E. limbic cortex

14. The floccular lobe of the cerebellum controls which of the following?
    A. gross limb movement
    B. fine coordinated hand movement
    C. eye movement
    D. dream sequences
    E. The function of the floccular lobe is currently unknown.

15. Which of the following is *not* true of the somatosensory system?
    A. Unlike other sensory systems, it is distributed across the entire body.
    B. It utilizes dendrites that respond to chemical stimulation.
    C. It utilizes dendrites that respond to mechanical stimulation.
    D. Animals adapt relatively easily to loss of all somatosensation.
    E. It includes sensory systems utilized for both balance and movement.

16. Which of the following describes the speed with which pain signals are sent compared with the speed with which touch and pressure signals are sent?
    A. Pain signals move approximately 10 times faster.
    B. Pain signals move approximately twice as fast.
    C. Pain signals move slightly faster.
    D. Both types of signals utilize the same dendrites and move at the same speed.
    E. Pain signals generally move slower.

17. When a doctor evokes a knee-jerk response from a patient by tapping the patellar tendon, the doctor is in fact stimulating which of the following?
    A. nocioceptive receptors
    B. hapsis receptors
    C. monosynaptic reflex
    D. pain gate response
    E. slowly adapting receptors

18. According to the pain gate theory, which of the following could be used to reduce a pain sensation?
    A. acupuncture
    B. electrical stimulation of some brain sites
    C. rubbing the area near the painful sensation
    D. giving an endogenous opiate such as morphine
    E. All of these could be used to reduce pain.

19. Which of the following is *not* considered a function of the vestibular senses?
    A. providing information about body position
    B. providing information about body temperature
    C. providing information about direction of body movement
    D. providing information about speed of body movement
    E. None of these are functions of the vestibular system.

20. Damage to the primary and secondary somatosensory cortex has *not* been shown to result in which of the following?
    A. reorganization of body regions represented by cortex surrounding the damage
    B. paralysis of the affected body part
    C. impaired motor function of the affected body part
    D. loss of sensation from the region associated with the damaged region
    E. apraxia

Short-Answer Questions

Answer each of the following questions with a brief but complete written answer based on information from your text.

1. Give a brief explanation of what is meant by hierarchical organization. Use control of movement as an example of this type of organization.

2. The brainstem is sometimes said to be a structure that controls "species-typical" behavior. Give several examples of such behavior with particular reference to studies conducted by Walter Hess in the 1950s.

3. Wilder Penfield found that brief electrical stimulation of motor cortex regions could evoke simple motor movements in corresponding body parts. More recently, researchers have expanded on those early findings by using longer duration electrical stimulation of these regions. Briefly describe what types of behaviors are evoked with longer duration of electrical stimulation.

4. The basal ganglia has been implicated in both hyperkinetic and hypokinetic disorders. Briefly describe what this tells us about the function of the basal ganglia.

5. Briefly characterize Tourette's syndrome. Include in your description behavioral features of the disorder and the underlying cause.

6. Briefly describe the types of symptoms you would expect to observe in an individual who suffers from cerebellar damage.

7. Briefly describe the difference between hairy skin and glabrous skin both in terms of location and in terms of physical features and function.

8. Briefly explain the concept of referred pain. Use for an example the pain felt by a heart attack victim.

9. Briefly explain why the homunculus for the motor cortex looks so similar to the homunculus for the somatosensory cortex.

10. Briefly describe what is meant by the term *apraxia*. Give an example of an apraxia and a possible cause of this dysfunction.

Matching Questions

Complete each of the following matching questions based on information from your text.

1. Match the following motor structures with their appropriate feature or description.

A. Caudate putamen	___ "Volume control" for movement
B. Prefrontal cortex	___ Planning movement
C. Primary motor cortex	___ Monosynaptic reflexes
D. Spinal cord	___ Executing movement
E. Cerebellum	___ Regulating posture

2. Match the following motor disorder with the associated damaged or dysfunctional region.

A. Cerebral palsy	___ Somatosensory cortex
B. Ataxia	___ Brainstem
C. Tourette's syndrome	___ Spinal cord
D. Disrupted eye movement and posture	___ Cerebellum
E. Paraplegia	___ Basal ganglia

3. Match the following somatosensory systems with the appropriate feature or description.

	___ Utilizes free nerve endings that release chemicals
A. Hapsis	___ Encapsulated nerve endings monitoring tendons
B. Nociception	___ Utilizes small unmyelinated fiber
C. Proprioception	___ Responds best to pressure
	___ Processes information on body position

4. Number from first (1) to last (5) the structures utilized in the dorsal spinothalamic tract to transmit information about body position.

___ Dorsal column nuclei
___ Ventrolateral thalamus
___ Muscle stretch receptors
___ Medial lemniscus
___ Somatosensory cortex

5. Match the following components of the vestibular system with their appropriate description.

A. Otoliths	___ May send excitatory or inhibitory signals
B. Semicircular canals	___ Senses changes in 3 planes of different orientations
C. Endolymph	___ Small crystals of calcium carbonate
D. Otoconia	___ Fluid filling the semicircular canals
E. Hair cells	___ Consists of the ultricle and saccule

Diagrams

1. In the diagram below identify the approximate location of the following regions associated with movement and somatosensation: Primary motor cortex, Primary somatosensory cortex, Cerebellum, Caudate putamen, Substantia nigra

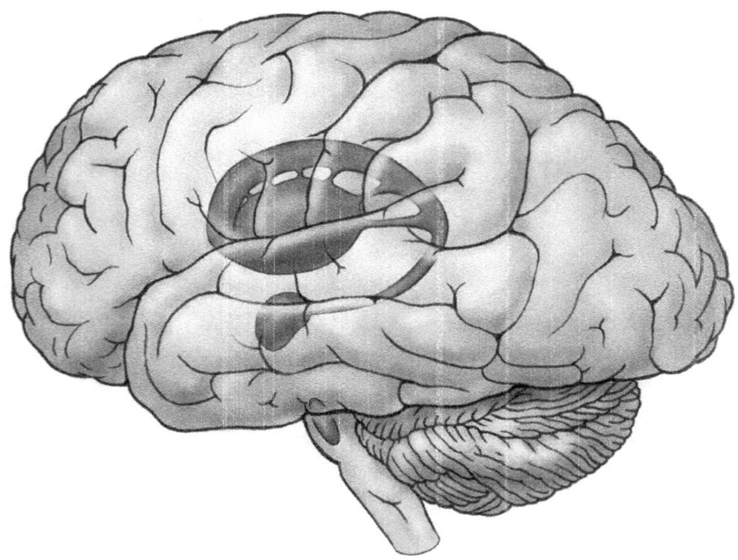

2. Draw a homunculus that approximately represents the relative volume of neural tissue associated with movement and somatosensation in a human.

3. Indicate on the diagram below the following structures: Caudate nucleus, Putamen, Globus paladus, Subthalamic nucleus, Substantia nigra.

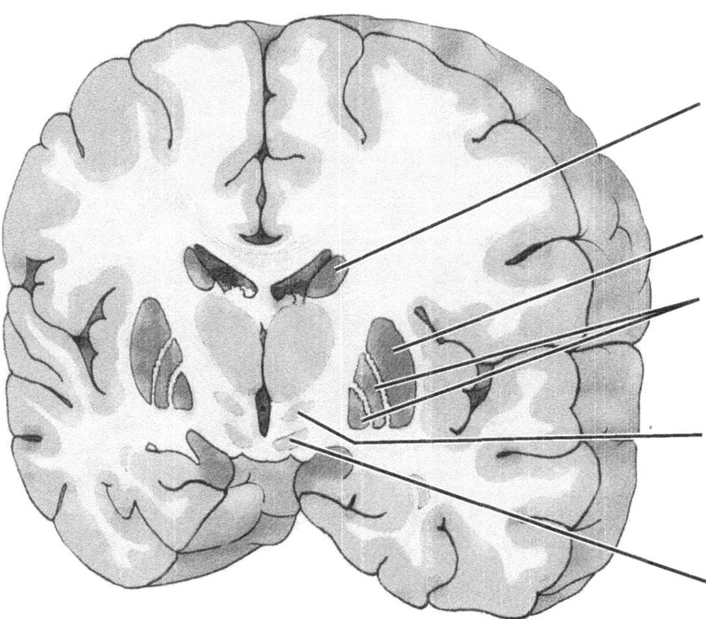

4. The diagram below is useful for explaining the gate theory of pain. Complete this diagram by indicating where there are question marks (I?) whether the input is excitatory or inhibitory. Also indicate at questions marks (A?) whether axons are large or small and whether they are myelinated or unmyelinated.

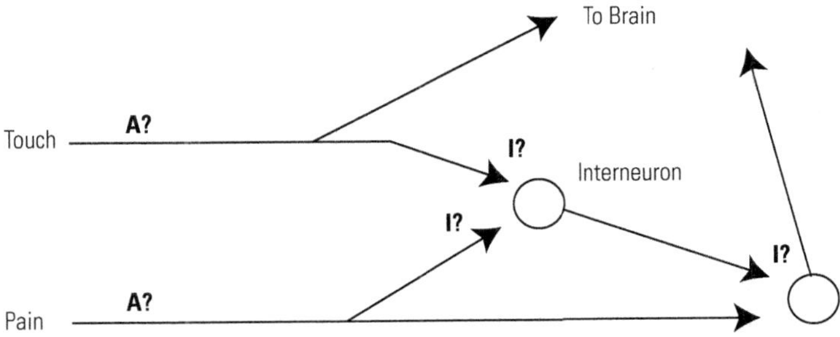

5. Identify the following regions on the diagram of the vestibular system structures shown below: Utricle, Saccule, Neural fibers, Semicircular canals.

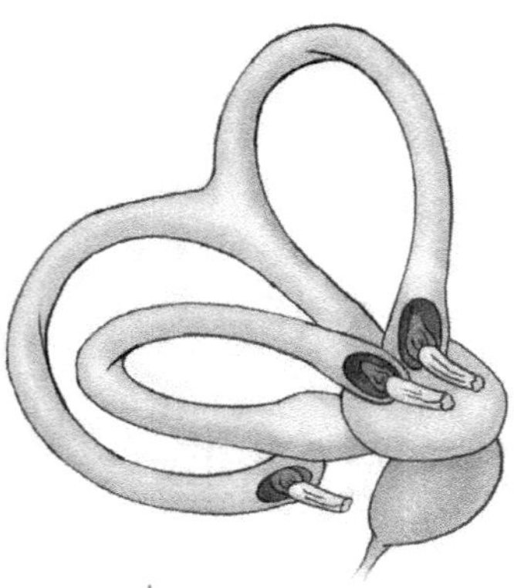

The Web

Consider using the following Web sites for additional information on some of the topics from this chapter:

1. The Christopher Reeve Paralysis Foundation: www.christopherreeve.org/
2. National Spinal Cord Injury Association: www.spinalcord.org/
3. Tourette's Syndrome Association: www.tsa-usa.org/
4. National Ataxia Foundation: www.ataxia.org/
5. United Cerebral Palsy: www.ucp.org/

CROSSWORD PUZZLE

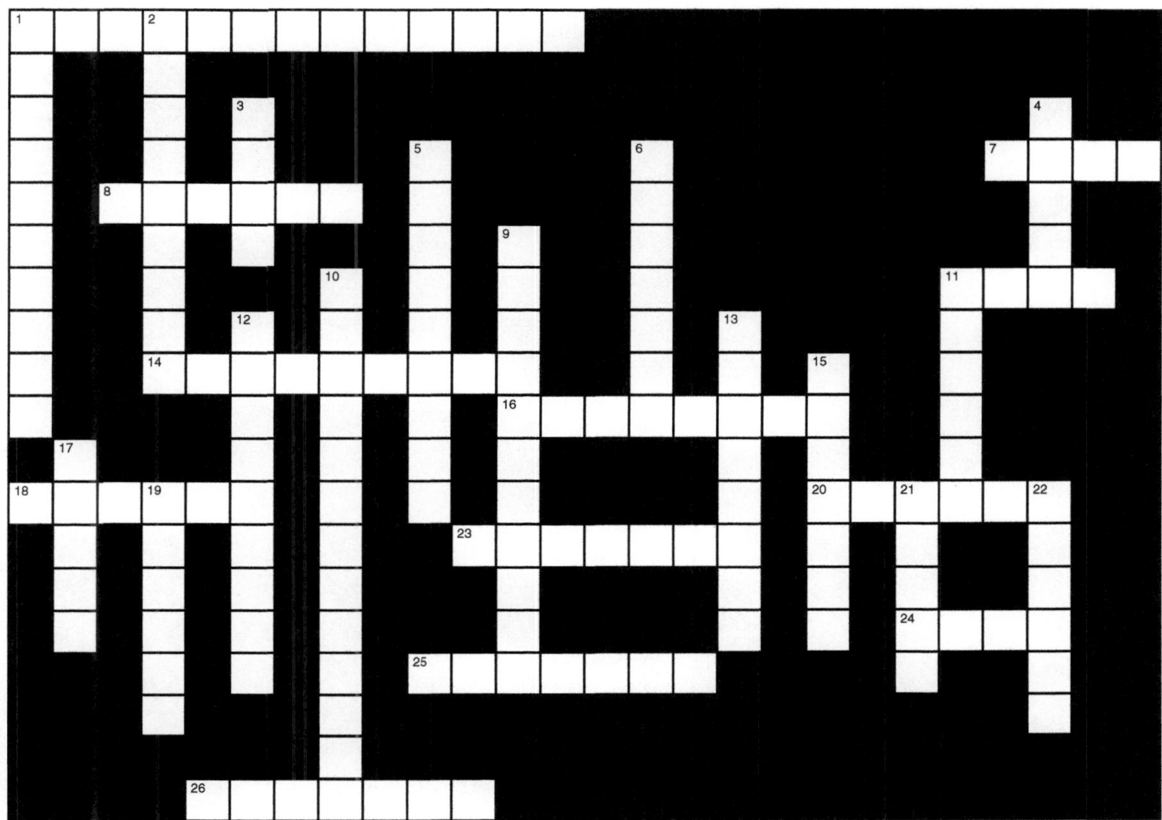

Across

1. To the bottom on the side, as in _____ thalamus
7. Somatosensation information enters the spinal cord via the dorsal _____
8. In the middle, as in _____ lemniscus
11. Some reflexes are found in the spinal _____
14. Fancy word for a basic movement pattern
16. Small crystals of calcium carbonate in the vestibular system
18. _____ root ganglion, where sensory information enters the spinal cord
20. Fancy term for sense of pressure or touch
23. This and a saccule make up 26 across
24. May be hairy or glabrous
25. A collection of cells, as in basal _____
26. One of the organs that makes up the vestibular system

Down

1. Sensory system that helps you keep your balance
2. Syndrome that includes tics and vocalizations
3. Simple term for nocioception
4. _____ cortex in the frontal lobe mediates movement
5. Lumpy cells on the ventral surface of the brainstem involved in movement
6. Hypo_____ means less movement
9. The opposite of evolution, seen after brain damage
10. Meaning from the cortex to the spinal cord
11. They're semicircular in the vestibular system
12. Fluid that fills 11 down
13. He did brain stimulation studies on his friends
15. He suggested motor sequences
17. Ventral _____ of spinal cords contain motor neurons and interneurons
19. Vesticular, motor, or somatosensory
21. Motor disorder, cerebral _____
22. Type of reflex seen in knee-jerk response

CHAPTER

12 What Causes Emotional and Motivated Behavior?

CHAPTER SUMMARY

The focus of this chapter is analysis of *motivated* (purposeful) *behavior*. In attempting to identify a cause of behaviors, early researchers found that both animals and humans appear to engage in some behaviors with no obvious function but simply to stimulate the brain in a rewarding fashion. Later researchers developed a "drive" theory, concluding that behaviors were driven by some form of internal energy, and that the vigor and duration of engaging in a behavior reflected the amount of energy stored for that specific action. Most recently researchers have begun to establish a neurochemical basis for many aspects of behavior. Sexual behavior, for example, correlates strongly with levels of circulating hormones. Electrical brain stimulation of areas releasing, or affected by, these mediating neurochemicals can evoke specific behaviors. A neural basis for such rewarding behaviors may be an important evolutionary adaptation to enhance species-specific behaviors that increase chances for survival. Bird killing by cats provides a good example of this explanation. The term *innate releasing mechanism* (*IRM*) is used to describe neural control of adaptive behaviors that appear without the influence of learning.

The idea that a rewarding neural circuitry underlies the generation of behaviors suggests that modifying some behaviors may be difficult. More specifically, some mechanism must exist for stimuli, either internal or external, to influence the internal reward circuitry. For example, internal levels of circulating hormones are highly correlated to sexual behavior. Externally, pheromones, a class of odorants sensed by the olfactory system, are capable of directly affecting the physiology or behavior of the recipient animal. Pheromones are also a unique class of odors, in that they are not detected by the general olfactory system, but by a specialized system that includes the *vomeronasal organ*. Vomeronasal organ activation is generally associated with changes in reproductive behaviors. Although the *gustatory system* is less directly linked to behavioral changes, some tastes do appear to activate the reward circuitry. There is little question that tastes significantly influence feeding behavior, which is likely linked to the neural connections of the gustatory nerve to the hypothalamus.

While IRMs may influence many behaviors, not all behaviors are controlled by IRMs. In addition, IRM-mediated behaviors are capable of modification, particularly using environmental stimuli that naturally affect the associated reward circuitry. *B.F.*

Skinner was among the first researchers to show *reinforcers* and punishment could modify many adaptive behaviors. However, substantiating the existence of a fairly rigid neural system, Skinner and others found that some behaviors could be modified only with specific associated stimuli. For example, *taste aversion learning* quickly and effectively modifies eating behavior with an aversive gastrointestinal stimulus. Electrical shocks and loud noises are far less effective at modifying eating behaviors, presumably because these aversive stimuli do not effectively disrupt the eating reward system driving behavior. The term *preparedness* is used to describe the phenomenon of certain stimuli being more easily associated with specific behaviors.

The term *motivation* is commonly used to describe purposeful behavior in animals. Motivation does not, however, describe something tangible found in the brain. Rather, it is a term for inferences made about why we engage in behaviors. For example, *regulatory behaviors* like feeding and drinking are motivated by *homeostatic mechanisms.* Homeostatic mechanisms are activated when a physiological state is altered from a certain set point. A glucostat mechanism indicates when blood glucose is low, motivating eating behavior, or when glucose is high, causing feeding to cease. Regulatory behaviors are mediated by the hypothalamus. *Nonregulatory behaviors* include all other behavior and are mediated by a variety of forebrain structures, most prominently the frontal lobe.

The hypothalamus influences behavior through both endocrine responses and autonomic nervous system activation. The *medial forebrain bundle (MFB)* is a dopamine-containing fiber tract that makes up the major hypothalamic pathway used in motivated behavior. The hypothalamus also mediates pituitary gland function. Endocrine responses are in turn initiated by release of pituitary hormones. The pituitary gland is actually only half endocrine gland (the anterior half). The posterior pituitary is composed of neural tissue directly influenced by hypothalamic connections. The anterior and posterior pituitary regions release different combinations of hormones affecting many different areas of the body and subsequently affecting many different behaviors. Hypothalamic function is mediated by three mechanisms. First, a *feedback mechanism* directly monitors the blood levels of hormones, decreasing and increasing hypothalamic function as hormone levels rise and fall around the homeostatic set point. Second, the hypothalamus receives direct input from the limbic system and the frontal lobe. Third, hypothalamic neurons undergo structural and biochemical changes in response to experience. Direct electrical stimulation of the hypothalamus produces a variety of behaviors depending on the location of the electrode. Three features of these behaviors provide insight into the types of behaviors controlled by the hypothalamus. First, stimulated behaviors are smooth, well-integrated, goal-directed behaviors. Second, these behaviors are related to survival (e.g., eating, drinking). Third, animals appear to find stimulation of these behaviors pleasant.

The limbic system includes several structures that comprise a primitive cortex sometimes called a limbic lobe. James Papez was among the first to speculate a contribution of the limbic lobe to emotions, noting dramatic emotional changes in people who suffered from rabies damage to neural structures in this lobe. The *hippocampus*, *amygdala*, and prefrontal cortex are considered limbic structures. Each of these structures has extensive neural connections to the hypothalamus. Stimulation of the amygdala evokes a fear response in animals, while lesions produce changes in feeding and hypersexuality. Abnormalities in dopamine projections from the prefrontal cortex have been implicated in the emotional blunting of schizophrenia. Damage to the frontal lobe may result in difficulty focusing on tasks, making individuals easily distractible. In this regard, the frontal lobe may be considered a structure mediating selection of behaviors.

Emotions are expressed through both physiological changes and motor behaviors. Like motivated behaviors, the hypothalamus is important for mediating emotional responses. In addition, the amygdala and cortical regions (primarily in the frontal lobe) are also involved. Determining how emotional responses are stimulated has long been debated. The *James-Lange theory*, or *somatic marker hypothesis* (a more recent variation), suggests emotions are not set responses, but rather interpretive responses to both autonomic changes and the stimuli evoking the physical response. This theory is bolstered in part by research showing that individuals who suffer from high spinal-cord injury (and thus lose much sensation of autonomic changes) report feeling weaker emotional responses to environmental stimuli than individuals with low spinal-cord injury. The amygdala appears to play an especially important role in emotional responses to fear-provoking stimuli. When this structure is bilaterally damaged, animals exhibit a marked decrease in aggression and fear responses known as *Klüver-Bucy syndrome*.

Emphasizing the role of the frontal cortex in emotional responses, researchers cite findings from patients undergoing frontal *lobotomy*, a form of *psychosurgery*. Such surgery results in patients who exhibit decreased emotion in facial expression and decreased prosody (emotional inflection in speech). These patients also have difficulty interpreting emotions expressed by others. It is believed the cortical tissue from the frontal lobe is responsible for providing a cognitive interpretation of autonomic responses.

Knowing that brain structures underlie emotions, it is not surprising that imbalances in brain chemistry and function can result in emotional disorders. *Depression* affects nearly 10 percent of the population. There is a genetic component to depression that likely results in dysfunction of neurotransmitter transmission in structures associated with emotions. The most effective antidepressant drugs increase noradrenaline and serotonin transmission, implicating these two neurotransmitters as primarily responsible for this disorder. *Anxiety disorders* are the most prevalent of all psychiatric disorders, affecting somewhere between 15 and 30 percent of the population. The most effective *anxiolytic* drugs are the *benzodiazepines* that act as agonists on GABA receptor sites. Anxiety is a useful emotion for survival, reducing contact with feared or dangerous stimuli. Anxiety disorders are thought to be an overactivity of this normally useful emotion.

Feeding is a regulatory behavior influenced by the digestive system, the hypothalamus, and by cognitive factors. The digestive system provides glucose for brain function. When glucose levels are reduced the brain initiates feeding behavior. When food is introduced into the lower digestive tract (beginning of the intestine) *cholecystokinin (CCK)* is released. CCK signals the hypothalamus, which in turn evokes a sensation of diminished hunger. Lesions of the lateral hypothalamus or fibers passing through this region produce *aphagia*, while lesions of the ventromedial hypothalamus produce *hyperaphagia*. These findings suggest an essential role for the hypothalamus in mediating initiation and cessation of feeding behavior. In particular, neurons in the hypothalamus that sense glucose levels (glucostatic neurons) and lipids (lipostatic neurons) may monitor blood concentrations and regulate behaviors to maintain homeostatic levels of these substances. Cognitive factors such as thinking about food or associations made to food odors may also influence eating behavior. The inferior frontal cortex receives direct input from olfactory bulbs, suggesting a role for this brain region in cognitive influences in feeding. The amygdala also appears to be involved in various aspects cognitive of eating, including the development of food preferences and aversions.

Drinking is a regulatory behavior that may be evoked by osmotic or hypovolemic thirst. *Osmotic thirst* is stimulated when a high concentration of salt in the system draws fluids from the cells into extracellular space. Cells surrounding the third ventricle act as detectors for these osmotic changes. *Hypovolemic thirst* is stimulated when fluid volume (generally blood volume) decreases. When this happens the kidneys detect the blood pressure decrease and send a hormonal signal to the hypothalamus to increase fluid consumption. Hypovolemic thirst differs from osmotic thirst in that consumption of both water and solutes (e.g., salt) is preferred over consumption of water only.

Nonregulatory, like regulatory, motivated behaviors are strongly influenced by hormones and hypothalamic function. Sexual behavior provides a good model for assessing these influences. Gonadal hormones (i.e., androgens) may have *organizing effects*, such as influencing the anatomical makeup of structures during fetal development. Interestingly, testosterone has its masculinizing effects on neural structures only after it is converted to estradiol by intracellular aromatase. Females are protected from the masculinizing effects of their own estradiol by a liver enzyme called *alpha fetoprotein*. Neural *sexual dimorphism* is most evident in the hypothalamic region of the *medial preoptic area*, which is approximately five times larger in males than in females. Gonadal hormones also have *activational effects*, evoking behaviors such as sexual response in female rats. The *lordosis* response shown by sexually receptive female rats, for example, is mediated by the ventromedial hypothalamus but only exhibited when estrogen and progesterone levels are adequate to stimulate this region. A clear link between neuroanatomy and sexual orientation has not been established, but some evidence suggests that structural differences exist between some heterosexual and homosexual men.

The opposite sensation from anxiety is reward. Reward is a useful emotion for inducing and maintaining contact with useful stimuli (such as food and mates). In the 1950s it was found that animals would engage in *intracranial self-stimulation* of certain brain regions. Among the most effective areas for *brain stimulation reward* is the *medial forebrain bundle*, also known as the *mesolimbic dopamine pathway*, terminating in the *nucleus accumbens*. One common feature of all the rewarding pathways for self-stimulation is the existence of dopamine-containing fibers. It is also interesting to note that recreational drugs that are addictive and/or frequently abused stimulate pathways containing dopamine fibers. It short, it appears these reward pathways evoke strong positive emotions that may be equally associated with either useful stimuli (food, mates) or addictive drugs.

KEY TERMS

The following is a list of important terms introduced in Chapter 12. Give the definition of each term in the space provided.

Evolutionary and Environmental Influences

Motivation

Emotion

Innate releasing mechanism (IRM)

Reinforcers

Taste aversion learning

Olfaction

Pheromones

Gustation

Preparedness

Neuroanatomy of Motivated Behavior

Regulatory behaviors

Homeostatic mechanism

Nonregulatory behaviors

Hypothalamus

Pituitary gland

Medial forebrain bundle (MFB)

Releasing hormones

Feedback loops

The Limbic System

Hippocampus

Papez circuit

Amygdala

Frontal lobe

Prefrontal cortex

Stimulating Emotion

James-Lange theory

Somatic marker hypothesis

Klüver-Bucy syndrome

Psychosurgery

Frontal leukotomy

Depression

Anxiety disorders

Anxiolytic drugs

Benzodiazepines

Generalized anxiety disorder

Phobias

Panic disorder

Control of Regulatory Behaviors

Obesity

Anorexia nervosa

Cholecystokinin (CCK)

Aphagia

Hyperphagia

Osmotic thirst

Water intoxication

Hypovolemic thirst

Control of Nonregulatory Behavior

Sexual behavior

Organizing effect

Gonadal hormones

Activating effect

Sexual dimorphism

Alpha fetoprotein

Preoptic area

Lordosis

Medial preoptic area (POA)

Androgen insensitivity syndrome

Sexual orientation

Sexual identity

Congenital adrenal hyperplasia

Androgenital syndrome

Reward

Intracranial self-stimulation

Brain stimulation reward

Mesolimbic dopamine pathway

Nucleus accumbens

Incentive

Reward

KEY NAMES

The following is a list of important names introduced in Chapter 12. Explain the importance of each person in the space provided.

B. F. Skinner

John Garcia

James Papez

PRACTICE TEST

Multiple-Choice Questions

Answer each of the following multiple-choice questions with the best possible answer based on information from your text.

1. What happened in the 1950s when Hebb and colleagues allowed subjects to exist in a stimulation-free environment where they did not have to engage in any motivated behaviors?
   A. Subjects generally became very bored in less than 24 hours before asking to discontinue.
   B. Subjects slept for an average of 3 days before asking to discontinue.
   C. Subjects suffered severe depression for an average of 3 days before asking to discontinue.
   D. Subjects suffered hallucination during an average of 3 days before asking to discontinue.
   E. Subjects enjoyed themselves for an average of 7 days before the study was discontinued.

2. As described in the text, many animals engage in the behavior of flehmen. In doing so, they draw volatile chemicals through their mouth in an attempt to stimulate which of the following?
   A. olfactory receptors
   B. taste buds
   C. capillary beds in the lungs
   D. the vomeronasal organ
   E. osmotic receptors

3. B. F. Skinner was among the earliest and most adamant proponents of the theory that motivated behaviors, including avoidance behaviors such as phobias, could be explained by which of the following?
   A. genetics
   B. reinforcement history
   C. neurotransmitter imbalance
   D. hormones
   E. evolution

4. Which of the following would *not* be considered a behavior mediated by a homeostatic mechanism?
   A. eating
   B. drinking pure water
   C. drinking water containing salts
   D. putting on a sweater
   E. engaging in sexual behavior

5. Which of the following is *not* implicated as a major contributor in the control of non-regulatory behaviors?
   A. hypothalamus
   B. pituitary gland
   C. frontal lobe
   D. occipital lobe
   E. the limbic system

6. Which of the following neurotransmitters is found in high concentrations in the medial forebrain bundle and considered a major chemical influencing motivated behaviors?
   A. dopamine
   B. acetylcholine
   C. epinephrine
   D. serotonin
   E. endorphins

7. The frontal lobe is thought to play an important role in selecting behaviors to be exhibited. Damage to the frontal lobe often results in which of the following?
   A. inability to show any emotion
   B. inappropriate expression of emotions
   C. chronic depression
   D. chronic anxiety
   E. None of the answers is correct.

8. According to the James-Lange theory of emotions, your experience of emotions is based on which of the following?
   A. stimulation of the spinal cord
   B. autonomic nervous system response
   C. cognitive function
   D. cognitive interpretation of an autonomic nervous system response
   E. an autonomic nervous system response to cognitive functioning

9. Which of the following is *not* a symptom of Klüver-Bucy syndrome?
   A. a tendency to examine objects by mouth
   B. an increase in homosexual behavior
   C. lack of fear
   D. aphasia marked by picky eating habits
   E. extreme tameness

10. Schizophrenia is thought to result in part from dysfunctional input from the frontal lobe to the hypothalamus. Which of the following is a feature shared by schizophrenics and patients who have undergone frontal lobotomy?
    A. lack of facial expression
    B. hallucinations
    C. delusions
    D. manic episodes
    E. All of the answers are correct.

11. Which of the following is *not* true of the emotional disorder of depression?
    A. There is a genetic component.
    B. Serotonin is implicated as a primary neurotransmitter in this disorder.
    C. It affects approximately 35 percent of the population.
    D. Approximately 70 percent of people reporting depression respond to drug treatment.
    E. It is among the most common psychologically disruptive disorders in the world.

12. Anxiety disorders are thought to be a result of which of the following?
    A. abnormal dopamine-receptor response in the hypothalamus
    B. abnormal dopamine-receptor response in the amygdala
    C. abnormal GABA-receptor response in the hypothalamus
    D. abnormal GABA-receptor response in the amygdala
    E. abnormal frontal-lobe function

13. What is the primary difference between a phobia and a panic attack?
    A. Panic attacks involve a clearly dreaded object or situation.
    B. Phobias involve a clearly dreaded object or situation.
    C. Panic attacks are not considered an anxiety disorder.
    D. Phobias are not considered an anxiety disorder.
    E. There is no difference between the two; they are different names for the same disorder.

14. Which of the following organs contributes most to maintaining short-term homeostatic balance of blood glucose?
    A. kidneys
    B. liver
    C. stomach
    D. small intestine
    E. large intestine

15. When would you expect blood concentrations of cholecystokinin (CCK) to rise?
    A. one hour prior to a meal
    B. at the onset of a meal
    C. near the end of a meal
    D. one hour after a meal
    E. approximately the middle of the night

16. Which of the following would be most likely to induce osmotic thirst?
    A. loss of blood due to a severe knife wound
    B. loss of blood due to a minor bloody nose
    C. ingestion of distilled water
    D. ingestion of a slice of watermelon
    E. ingestion of a bag of potato chips

17. Which of the following is responsible for masculinizing sexually dimorphic neurons in the male brain during fetal development?
    A. testosterone
    B. estrogen
    C. testosterone that is converted into estrogen
    D. estrogen that is converted into testosterone
    E. None of the answers is correct.

18. In what structure are the sexually dimorphic organizational effects of gonadal hormones most apparent?
    A. hippocampus
    B. hypothalamus
    C. amygdala
    D. frontal lobe
    E. cerebellum

19. Research on hypothalamus dimorphism comparing homosexual males, heterosexual males, and females suggests that the hypothalamus of homosexual males:
    A. is most similar to that of heterosexual males.
    B. is most similar to that of females.
    C. is different from both heterosexual males and females.
    D. is similar to females only in homosexual males who exhibit overt feminine behavior.
    E. is similar to females only in homosexual males who exhibit overt masculine behaviors.

20. The mesolimbic region is considered a reward pathway that utilizes which of the following?
    A. dopamine
    B. GABA
    C. serotonin
    D. testosterone
    E. estrogen

Short-Answer Questions

Answer each of the following questions with a brief but complete written answer based on information from your text.

1. What is an innate releasing mechanism (IRM)? Give an example of an IRM in humans.

2. Briefly describe taste aversion learning as demonstrated by John Garcia in wolves. Could taste aversion learning be established by pairing a particular food with a loud noise? Explain your answer.

3. Explain what is meant by a homeostatic mechanism. Use eating as an example of a behavior associated with a homeostatic mechanism.

4. Briefly describe the James-Lange theory of human emotional experiences.

5. Briefly explain the basis for the Klüver-Bucy syndrome and list at least three features of this syndrome.

6. As a form of psychosurgery the frontal lobotomy has a rich history. Describe the results from initial animal studies of lobotomies and give an example of why this surgery might have been performed on a human.

7. There are several distinct types of anxiety disorders. List three subtypes of anxiety and briefly describe characteristics of each type.

8. Give an example of an organizational and an activation effect of gonadal hormones.

9. Male rats with medial preoptic area (POA) lesions do not mate. Briefly describe how Barry Everitt showed that this decrease in mating behavior did not indicate a loss of sex motivation in these rats.

10. Briefly explain why a person would continue to engage in the self-destructive behavior of cigarette smoking when they know such behavior is not useful and may even be detrimental.

Matching Questions

Complete each of the following matching questions based on information from your text.

1. Match the following structure to the behavioral response you would expect to see if the structure were damaged or lesioned.

A. Amygdala	___ Lack of prosody
B. Lateral hypothalamus	___ Reduced fear response
C. Ventromedial hypothalamus	___ Hyperphagia
D. Frontal lobe	___ Disruption of male sexual behavior
E. Medial preoptic area	___ Aphagia behavior

2. Match the following disorders to the appropriate feature or characteristic.

	___ Treated with anxiolytic drugs
A. Schizophrenia	___ Reduced dopamine from frontal lobe
B. Depression	___ Disruption of serotonin and noradrenaline
C. Anxiety disorder	___ Overactivity of $GABA_A$ receptors
	___ Affects nearly 10 percent of the population

3. Match the following physiological change with the expected behavior or sensation.

A. Loss of blood	___ Satiety
B. Ingestion of salt	___ Fear response
C. Release of CCK	___ Hypovolemic thirst
D. Release of estrogen	___ Osmotic thirst
E. Stimulation of amygdala	___ Sexual receptivity

4. Match the following anxiety disorders with their appropriate symptom or characteristic.

	___ Persistent and unrealistic worries
A. Generalized anxiety disorder	___ Involve a clearly dreaded object or situation
B. Panic disorder	___ Recurrent attacks of intense terror
C. Phobia	___ Often leads to agoraphobia
	___ Most common type of anxiety disorder

5. Label each of the following as an organizational (O) or activational (A) effect of hormones.

___ Testosterone producing sexual dimorphism of preoptic area
___ Testosterone producing male sexual behavior
___ Estrogen and progesterone producing lordosis
___ Congenital adrenal hyperplasia causing an enlarged clitoris
___ Estrous cycle hormones increasing dendritic branching

Diagrams

1. Below is a frontal section through the hypothalamus. Indicate the approximate location of lesions that would produce the following: 1) Aphasia, 2) Hyperaphagia, 3) Disruption of lordosis response.

2. On the diagram below, indicate approximately how you would perform a lobotomy.

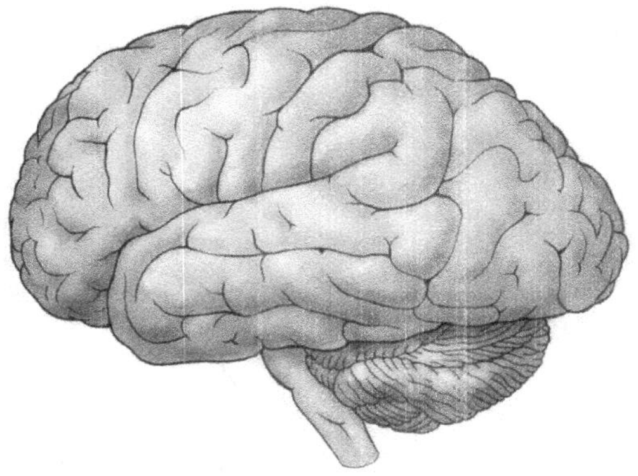

3. Below is an incomplete diagram of a pituitary gland connected to the hypothalamus. Complete the diagram by correctly subdividing the pituitary. Also indicate where hypothalamic neurons terminate in the pituitary gland.

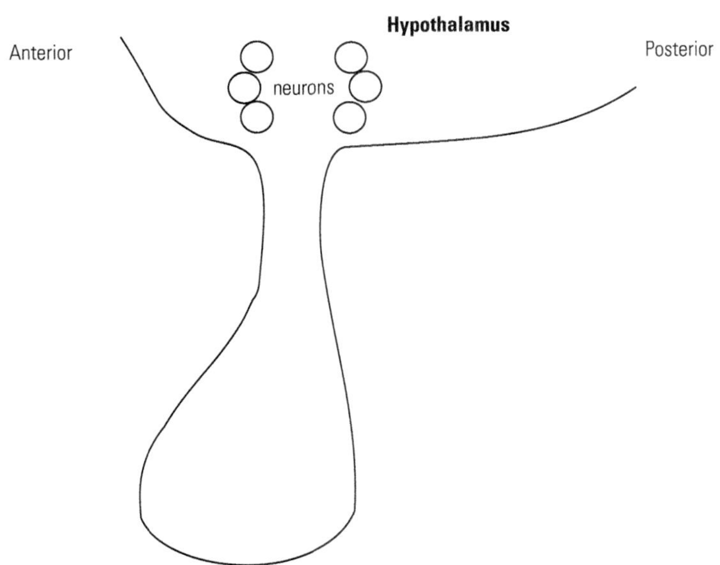

4. Below is a diagram of a cell in homeostatic balance. Alter this diagram to indicate a state that would induce osmotic thirst.

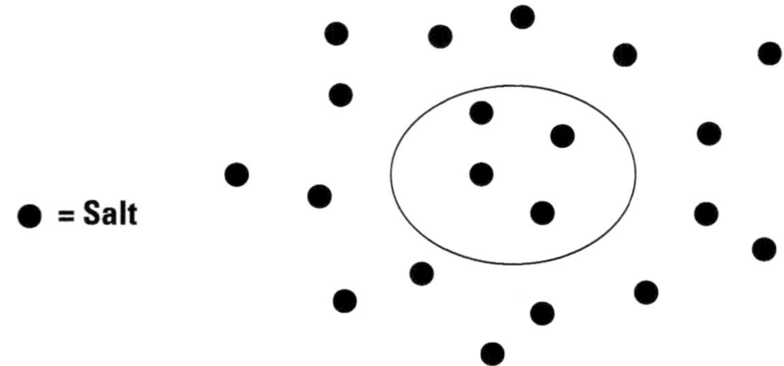

5. Indicate which of the neurons shown below would be masculinized during development.

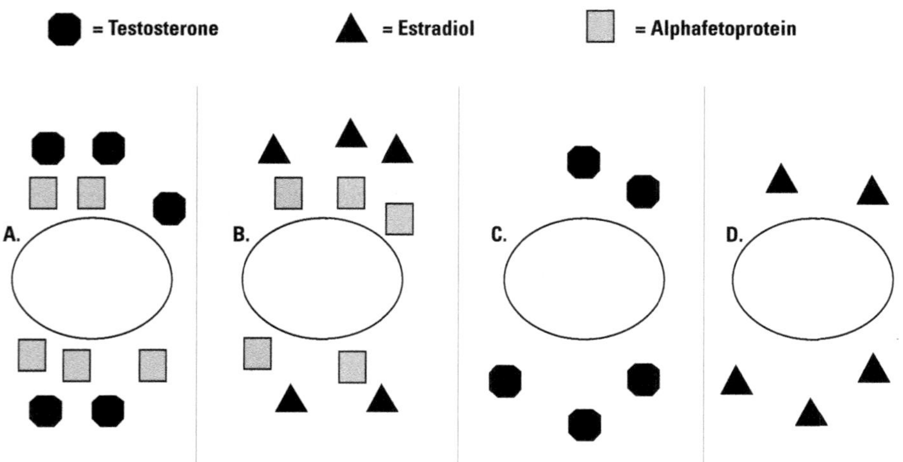

The Web

Consider using the following Web sites for additional information on some of the topics from this chapter:

1. American Anorexia Bulimia Association, Inc.: www.aabainc.org/
2. Obesity help and resources: www.obesityhelp.com
3. Anxiety Disorders Association of America: www.adaa.org/
4. Depression resource center: www.healingwell.com/depression/
5. Androgen Insensitivity Syndrome Support Group (AISSG): www.aissg.org

CROSSWORD PUZZLE

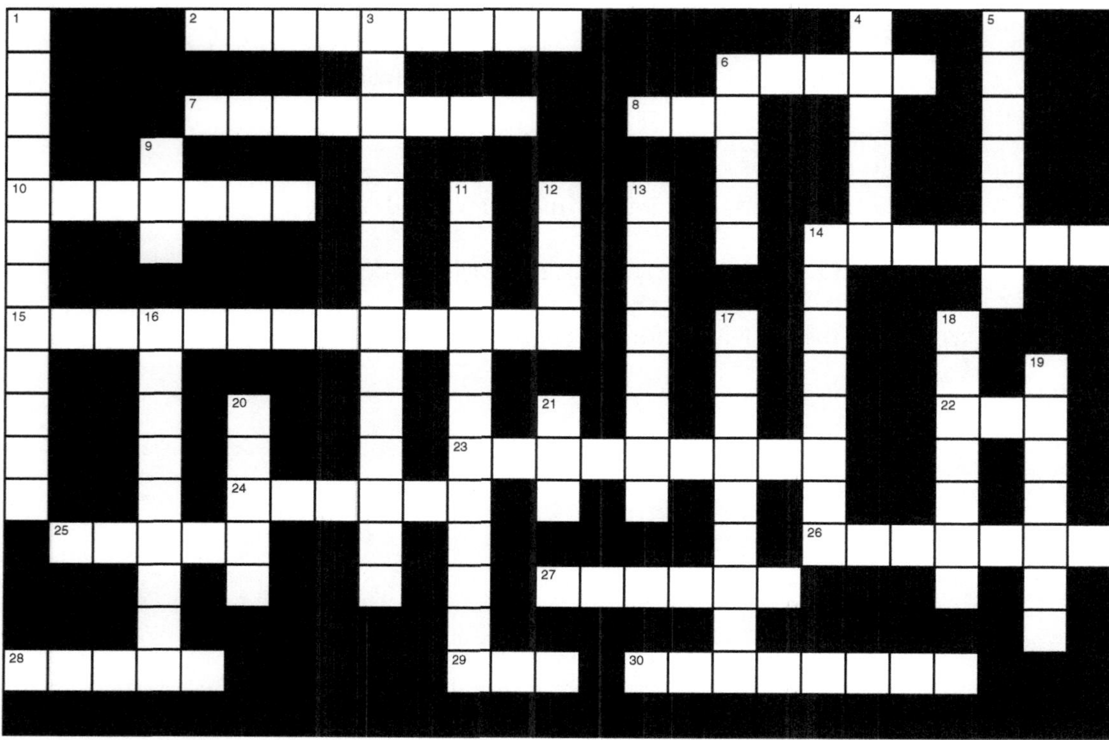

Across

2. The "F" in MFB
6. He proposed an emotional circuit after observing people with rabies
7. Process of remembering a behavior
8. 30 across; abbr.
10. Thirst induced by salt ingestion
14. Gland above the kidney, becomes active when you're stressed
15. A behavior that is not a regulatory one
22. Behavioral response to a fearful stimulus
23. The nucleus _____ is implicated in many theories of drug addiction
24. The "M" in MFB
25. Sensation that can enhance eating behavior
26. #1 name in pigeon learning research, maybe all learning research
27. He used taste aversion learning to keep wolves away from sheep
28. 2nd name in 20 down theory
29. An example of 15 across
30. Area responsible for male sex behavior; medial _____

Down

1. A syndrome that can masculinize women
3. Class of GABA agonists; e.g., valium
4. Something that encourages behavior
5. An example of a regulatory behavior
6. Disorder marked by high levels of anxiety
9. High level of negative emotions can make you do this
11. Primary subcortical structure associated with all types of motivated behavior
12. Lesions of the amygdala produce the Klüver-_____ syndrome
13. Surgery designed to disconnect a portion of the frontal cortex
14. Failure of a structure to develop
16. The "R" in IRM
17. Robinson and Berridge developed the _____ sensitization theory of drug addiction
18. Testosterone is an example of one
19. Unsettling emotion, treated with 3 down when it becomes chronic
20. With 28 across, they had a theory of emotions named after them
21. Cholecystokinin; abbr.

CHAPTER

13 Why Do We Sleep and Dream?

CHAPTER SUMMARY

Sleep appears to be a behavior that, like many other behaviors, helps animals adapt to their environment to maximize food acquisition and minimize energy loss. Humans are *diurnal* animals, meaning we are active and gather food during daylight and we are inactive and conserve energy at night. The phenomenon of our body systems actually slowing down at night to conserve energy and then speeding up during the day is called a *biological rhythm*. One very interesting question that may be asked is where such rhythms originate. The concept of a biological clock that keeps track of time and influences biological rhythms has been a theme of philosophy and research for nearly three centuries. Only recently, however, has extensive study begun to reveal details of the biological basis for these rhythms. Among the findings was that numerous cycles or *periods* of activity exist. In addition to *circadian* (around a day) cycles, research has shown *ultradian* (smaller than a day), *infradian* (between one day and one year), and *circannual* (around a year) rhythms. Of greatest interest to neuroscientists are behaviors mediated by circadian rhythms, including our sleep–wake cycle. Studying these cycles researchers determined that circadian rhythms are not entrained to an exact 24-hour period, but rather appear to respond to light and dark cues in the environment to reset the rhythm each day. When constant light is present, the time period generally extends beyond a 24-hour period, resulting in a *free-running rhythm* of about 25–27 hours for humans. Thus it appears that individuals have a personal rhythm period to which they would adhere without environmental light cues. However, with light cues all rhythms may be reset each day to approximately 24 hours. Such light cues are called *zeitgebers*, literally meaning "time-givers." One interesting behavioral phenomenon related to circadian rhythms is the effect of *jet lag* on travelers. Because our endogenous rhythms tend to be greater than 24 hours, travel from east to west (where daylight is gained) tends to be less disruptive than travel from west to east. In other words, west-to-east travel results in a greater discrepancy between our endogenous-rhythm period and the zeitgeber, making adjustments more difficult.

Of particular interest to researchers was the neural basis of the biological clock mediating rhythms. In the 1930s *Curt Richter* proposed that the biological clock acted as a *pacemaker* instructing other neural structures when to be active. After numerous attempts Richter determined that the *suprachiasmatic nucleus* (*SCN*) of the hypothalamus appears to act as the primary pacemaker for circadian rhythms. As the name suggests, this nucleus is located just dorsal from the optic chiasma and receives light input from the environment via the *retinohypothalamic tract*. Since Richter's research, neural pacemakers have also been located in the pineal gland and in the retina. However, the SCN is still considered the primary pacemaker for mediating most circadian rhythm. Not surprisingly, the SCN shows greater activity during the light portion of the light–dark cycle than during the dark portion. This activity continues even when the SCN is isolated from all neural input from other brain regions. SCN function also appears to be innate since animals raised in constant darkness show rhythmic activity. In addition, transplantation of SCN neurons from one animal to another will produce rhythms in the recipient animal that are consistent with those of the donor animal.

Although the SCN is the primary pacemaker, it is not itself responsible for producing behavior. Rather, the SCN appears to influence neural systems known as "slave" oscillators to produce changes in daily activities such as feeding, body temperature, sleeping, and so on. In addition to affecting daily activities, the SCN influences some annual behaviors (*circannual rhythms*). For example, animals that are seasonal breeders exhibit sexual activity only when changes in daylight length alter SCN mediation of associated hormones. Although gonadal hormones directly mediate sex behavior, these hormones may be regulated by *melatonin* secreted by the pineal gland in response to changes in light.

Sleeping and waking are among the most intriguing and thus the most studied behaviors. For the most part, we all generally sleep for several hours during each 24-hour period. One exception is *insomniacs* who are unable to sleep regularly or for adequate periods of time. To understand sleep disorders, we must first understand normal sleep patterns. Sleep is generally studied using physiological measures from a polygraph including EEG, EMG, and EOG to monitor changes in behavior. EEG has shown a reliable pattern of changes in brain waves as a person progresses from wakefulness into sleep. This pattern includes several overlapping stages of brain activity where electrical output shows progressively greater amplitude and lower frequency of brain-wave activity. These waves are called *beta* (15–30 Hz), *alpha* (7–11 Hz), and *delta* (1–3 Hz) *rhythms* and represent progression from wakefulness to deep sleep in respective order. Interestingly, after reaching the deepest stage of slow-wave delta-rhythm sleep, a person will then begin to cycle back up beta-rhythm brain activity in a state of *REM* sleep. During a typical night's sleep a person will cycle back and forth through deep sleep to REM approximately four to six times. As the night progresses, REM sleep is extended and *NREM* sleep is reduced. There is variability in individual sleep patterns, but most people exhibit a much larger proportion of REM sleep during infancy and childhood than in later adulthood. During slow-wave NREM sleep there are a variety of biological changes. Body temperature, heart rate, blood flow all decrease, while perspiring and release of growth hormone increase. We move about quite frequently, and in some individuals sleepwalking and night terrors may occur at this time. REM sleep is even more remarkable in terms of associated behaviors. The body becomes paralyzed with the exception of slight twitches. Body-temperature regulatory systems cease functioning, allowing body temperature to rise or fall with the surrounding environment. Brain-wave activity increases and the cognitive experience of dreaming occurs.

There are many theories of why we dream, including Freud's extensive overview published in the book *Interpretation of Dreams*. More contemporary theorists have proposed the *activation-synthesis hypothesis*, which suggests that the cortex is bombarded with neural signals during REM sleep and in response to these signals generates images and actions based on memory stores, often (but not always) using recent memories. The *evolutionary hypothesis of dreams* further suggests that dreams often include threatening elements of our environment as an adaptive means of enhancing ability to recognize and avoid such threats when awake.

As with dreaming, there are several theories as to why we sleep. Perhaps the earliest theory is that sleep is simply a passive process of brain inactivity occurring when environmental stimuli are reduced. However, research refutes this theory, showing that total lack of environmental stimuli actually tends to reduce time spent sleeping. Another theory is that sleep is a biological adaptation that allows conservation of energy during times when food is most difficult to obtain. Sleep may also be a restorative process, during which the body replenishes depleted chemicals, enzymes, and the like, used up during the wakeful period. However, some evidence suggests that total sleep deprivation, even for several days, does not appear to have lasting debilitating effects that might be expected if this theory were true. Sleep may also act to enhance memory storage, allowing the brain a period to solidify and organize experiences. Research has shown that sleep deprivation may reduce memory consolidation, and that dreams in particular may be responsible for aspects of memory storage.

The neural basis for sleep appears to be located primarily with the brainstem. The *reticular activating system* (*RAS*) includes the *reticulum*, and when activated stimulates *desynchronized* EEGs associated with wakefulness. Damage to the RAS results in a state of relatively permanent sleep commonly termed *coma*. Desynchronized EEGs are associated with acetylcholine released when alert but not moving, and serotonin released when moving. For sleep to occur, release of both of these neurotransmitters onto cortical neurons must be decreased. REM sleep appears to be mediated by a brainstem region just anterior to the cerebellum known as the peribrachial area. This structure sends cholinergic projections to the *medial pontine reticular formation* (*MPRF*) that in turn initiates PGO (pons, geniculate, occipital) waves associated with dreams.

Sleep disorders may occur during NREM and REM sleep. NREM disorders include *insomnia* (inability to sleep) and *narcolepsy* (falling asleep at inappropriate times). There are many factors that may contribute to sleep disorders. For example, approximately 35 percent of insomnia cases are associated with depression and anxiety. *Drug-dependency insomnia* results when drug-inducing sedatives create an imbalance in the normal chemical transmission associated with sleep, resulting in an inability to sleep without taking these drugs. Narcolepsy affects about 1 percent of the population and is distinguished from normal tiredness by its frequency and tendency to disrupt normal daily activities. One cause of narcolepsy is sleep apnea, an intermittent inability to breathe during periods of sleep, resulting in constant waking throughout the night. REM disorders include *sleep paralysis*, a fairly common condition in which an individual experiences paralysis normally associated with REM sleep as they are falling asleep or waking. *Cataplexy* is a rare form of sleep paralysis in which an individual loses muscle tone when fully awake, often during times of high emotional arousal. During a cataplexy episode, the individual may also experience *hypnogogic hallucinations* that are thought to result from stimulation of brain mechanisms normally activated during dreams. In contrast to sleep paralysis, *REM without atonia* is a condition in which individuals do not undergo paralysis during REM sleep. This condition can result in complex movements, often resulting in the person acting out the dream. Cataplexy and narcolepsy can be treated with amphetamines, whereas REM without atonia is treated with benzodiazepines that block REM sleep.

KEY TERMS

The following is a list of important terms introduced in Chapter 13. Give the definition of each term in the space provided.

A Clock for All Seasons

Diurnal

Biorhythms

Biological clocks

Biological rhythms

Period

Circannual rhythms

Circadian rhythms

Ultradian rhythms

Infradian rhythms

Free-running rhythms

Zeitgeber

Entrained

Jet lag

Neural Basis of the Biological Clock

Pacemaker

Suprachiasmatic nucleus

Slave oscillators

Retinohypothalamic tract

Melatonin

Sleep Stages and Dreaming

Electroencephalograph (EEG)

Electromyograph (EMG)

Electrooculograph (EOG)

Waking state

Drowsy state

Sleeping state

Dreaming state

Beta rhythm

Alpha rhythm

Delta rhythm

REM

NREM

Slow-wave sleep

Atonia

Interpretation of Dreams

Activation-synthesis hypothesis

Coping hypothesis

Basic rest–activity cycle (BRAC)

Microsleeps

Place cells

The Neural Basis of Sleep

Desychronized EEG

Reticulum

Reticular activating system (RAS)

Coma

Peribrachial area

Medial pontine reticular formation (MPRF)

Magnocellular nucleus

Sleep Disorders

Insomnia

Narcolepsy

Drug-dependency insomnia

Sleep apnea

Sleep paralysis

Cataplexy

Hypnogogic hallucinations

REM without atonia

Seasonal affective disorder (SAD)

Restless Legs Syndrome (RLS)

KEY NAMES

The following is a list of important names introduced in Chapter 13. Explain the importance of each person in the space provided.

Curt Richter

Sigmund Freud

PRACTICE TEST

Multiple-Choice Questions

Answer each of the following multiple-choice questions with the best possible answer based on information from your text.

1. Which of the following is an example of a circannual rhythm?
   A. migratory cycles
   B. eating behavior
   C. sleep behavior
   D. menstrual cycle activity
   E. All of the answers are correct.

2. In humans, the free-running nature of our biological clock is not usually apparent because with each day we reset the clock using environmental cues. Without cues to reset the clock, what would be the approximate period of a circadian rhythm?
   A. 12–14 hours
   B. 16–18 hours
   C. 20–22 hours
   D. 25–27 hours
   E. 29–31 hours

3. Which of the following is an example of a zeitgeber?
   A. food
   B. temperature
   C. light
   D. sexual behavior
   E. blood pressure

4. Within which of the following structures is the suprachiasmatic nucleus located?
   A. hippocampus
   B. hypothalamus
   C. brainstem
   D. cerebral cortex
   E. pituitary gland

5. Which of the following is true of the suprachiasmatic nucleus?
   A. Metabolic activity is higher during the day than at night.
   B. Neurons are more active during the day than at night.
   C. If all pathways into and out of this structure are severed, it maintains rhythmic activity.
   D. Individual cells from this structure each maintain a rhythmic activity.
   E. All of the answers are correct.

6. The suprachiasmatic nucleus does not actually drive behavior, but rather drives structures that in turn produce behavior. What is the term given to behavior-producing systems that are under the control of suprachiasmatic nucleus activity?
   A. slave oscillators
   B. entrainment systems
   C. circadian nuclei
   D. endocrine glands
   E. subordinate systems

7. Among the devices used for measuring sleep activity is the polygraph. Which of the following polygraph components is used when measuring sleep behavior?
   A. electroencephalograph
   B. electromyograph
   C. electrooculograph
   D. All of the answers are correct.
   E. None of the answers is correct.

8. Which of the following is true of EEG recordings as a person passes into progressively deeper stages of NREM sleep?
   A. wave frequency increases, wave amplitude increases
   B. wave frequency increases, wave amplitude decreases
   C. wave frequency decreases, wave amplitude decreases
   D. wave frequency decreases, wave amplitude increases
   E. Wave frequency and amplitude remain unchanged until the onset of REM sleep.

9. Which of the following is *not* a characteristic of slow-wave (NREM) sleep?
   A. increased body temperature
   B. increased perspiration
   C. decreased heart rate
   D. decreased blood flow
   E. increased growth hormone secretion

10. Which of the following is *not* a characteristic of REM sleep?
    A. penile erection
    B. paralysis of skeletal muscles
    C. twitching of fingers and toes
    D. loss of temperature regulation
    E. dreams generally last only a few seconds

11. According to the activation-synthesis hypothesis of dreams, a dream about your professor attacking you could result from which of the following?
    A. a high level of activation of the cortex
    B. a high level of output from the brainstem
    C. a recent emotional experience
    D. a recent encounter with that professor
    E. All of the answers are correct.

12. According to the coping hypothesis of dreams, a dream about your professor attacking you could result in which of the following?
    A. insomnia
    B. a fear of your professor
    C. a sexual attraction to your professor
    D. a desire to sit in the front of the classroom
    E. a desire to spend more time with your professor

13. In 1965 a student stayed awake for 260 hours (nearly 11 days). What lasting effect did this sleep deprivation have on this student?
    A. He now enters REM sleep much less than before the sleep deprivation.
    B. He now enters REM sleep much more than before the sleep deprivation.
    C. He now sleeps more hours per night than before the sleep deprivation.
    D. He now sleeps fewer hours per night than before the sleep deprivation.
    E. The sleep deprivation had no apparent lasting effects on the student.

14. One difficulty with sleep-deprivation studies is keeping subjects from engaging in ANY sleep. For example, a person can actually engage in a few moments of sleep, becoming less responsive to external stimuli and presumably gaining some benefits of sleep, while sitting or standing. What is the term researchers have given to these brief sleep periods?
    A. microsleeps
    B. cat naps
    C. power naps
    D. daytime deltas
    E. afternoon alphas

15. Supporting the theory of sleep as a function of memory storage, PET scans have shown that cortical areas that become active during task-acquisition trials in subjects also become active during which of the following?
    A. alpha-wave activity
    B. NREM sleep
    C. delta-wave activity
    D. REM sleep
    E. All of the answers are correct.

16. In general, it could be said that the neural structures responsible for initiating sleep are centered in which of the following?
    A. the hypothalamus
    B. the pituitary gland
    C. the brainstem
    D. the cortex
    E. the spinal cord

17. The region that appears to be responsible for sleep–wake behavior is known as the RAS. RAS is an abbreviation for which of the following?
    A. Region Associated with Sleep
    B. Reticular Activating System
    C. Rest and Sleep System
    D. Rest and Stimulation System
    E. Region of Alertness and Sleep

18. What two neurotransmitters released in the cortex are associated with desynchronized EEG waves?
    A. dopamine and serotonin
    B. dopamine and norepinephrine
    C. serotonin and norepinephrine
    D. acetylcholine and norepinephrine
    E. acetylcholine and serotonin

19. Which of the following is *not* true of cataplexy?
    A. It may be treated with amphetamine or Ritalin.
    B. Attacks may be triggered by excitement.
    C. Attacks may be accompanied by hallucinations.
    D. Attacks result in a loss of all muscle tone.
    E. Attacks generally occur when a person is drowsy and not alert.

20. Which of the following has been used to successfully treat REM without atonia?
    A. benzodiazepines
    B. muscle relaxants
    C. amphetamine
    D. ritalin
    E. alcohol

Short-Answer Questions

Answer each of the following questions with a brief but complete written answer based on information from your text.

1. Which type of travel tends to produce the more severe jet-lag symptoms, flying east to west or west to east? Briefly explain your answer.

2. The suprachiasmatic nucleus is a very descriptive name for a structure closely involved in biological rhythms. Briefly explain where this structure is located and why this location is important for circadian rhythms.

3. Briefly describe methods of experimental studies designed to show that rhythmic patterns of behavior are neither learned after birth nor entrained to the mother's rhythms during fetal development.

4. Briefly describe how the suprachiasmatic nucleus can mediate a circannual pattern of sexual behavior in a male animal. Include in your description the role of pineal gland and melatonin secretion.

5. Sigmund Freud suggested that many (perhaps most) dreams contained sexual content. Contemporary sleep researchers suggest only about 1 percent of dreams contain such content. What do these contemporary researchers see as the primary content of most dreams?

6. Briefly describe either the activation-synthesis hypothesis or the coping hypothesis of dreams.

7. Summarize very briefly the primary premise of each of the four following theories of sleep: 1) Sleep as a passive process; 2) Sleep as a biological adaptation; 3) Sleep as a restorative process; 4) Sleep as a means of enhancing memory storage.

8. REM sleep is thought to have some significance in human sleep behavior. This theory is strengthened by research investigating the effects of selective REM deprivation on subjects. Briefly describe two confirmed effects of REM deprivation that have been reported.

9. Briefly compare and contrast the sleep disorders of narcolepsy and cataplexy.

10. Briefly explain the relationship between narcolepsy and sleep apnea.

Matching Questions

Complete each of the following matching questions based on information from your text.

1. Match the following rhythms to their appropriate period.

A. Circannual rhythm
B. Circadian rhythm
C. Ultradian rhythm
D. Infradian rhythm

___ Around one year
___ Less than one day
___ Between one day and one year
___ Around one day

2. Match the following structures or regions to their appropriate feature or description.

A. Suprachiasmatic nucleus
B. Cerebral cortex
C. RAS
D. Median raphe
E. Brainstem

___ Damage to this may result in coma
___ PGO spikes originate here
___ Considered the location of the main biological clock
___ EEGs are recorded from this area
___ Contains serotonin neurons that project to neocortex

3. Indicate whether each of the following is associated with REM or NREM sleep.

A. _______ Delta rhythms
B. _______ Paralysis
C. _______ Night terrors
D. _______ Sleepwalking
E. _______ Loss of temperature-regulatory mechanism

4. Match the following sleep disorders with their appropriate descriptive feature.

A. Sleep apnea
B. Insomnia
C. Drug-induced insomnia
D. Hypnogogic hallucination
E. Night terrors

___ More common in people who are overweight
___ Anxiety and depression account for about 35 percent of cases
___ Results from tolerance development
___ May occur during a cataplexy attack
___ NREM disorder seen especially in children

5. Match the following drugs to the sleep disorder for which they are prescribed as treatment.

A. Benzodiazepines
B. Amphetamine
C. L-dopa
D. Light therapy

___ Seasonal affective disorder
___ Cataplexy
___ Restless Legs Syndrome
___ REM without atonia
___ Narcolepsy

Diagrams

1. The diagram below depicts waking periods of a normal subject over 8 days. Indicate how you would expect waking patterns to differ if this subject were deprived of light cues over that same time period (shaded bars indicate dark period, open bars indicate light period).

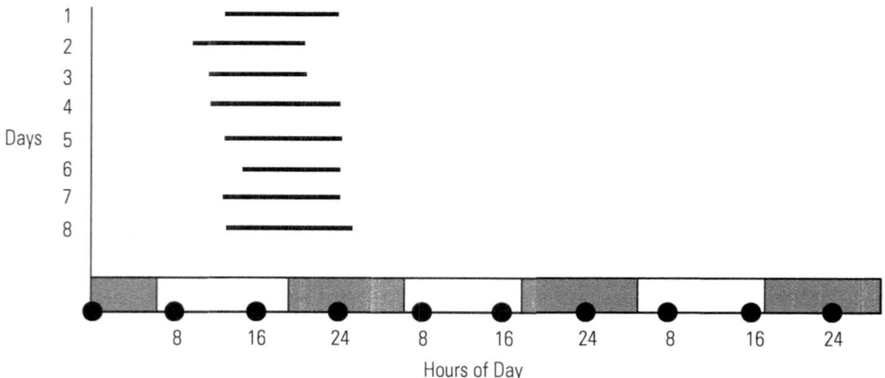

2. Which of the 4 travelers (indicated on the map below) would you expect to experience the greatest sensation of jet lag?

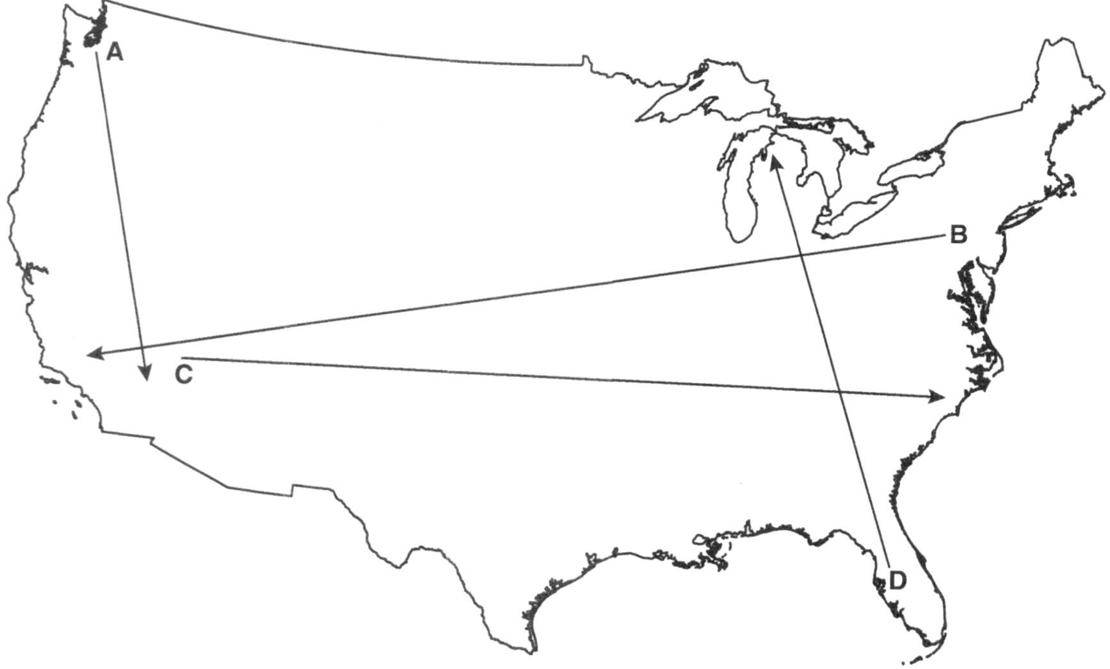

3. The line below depicts an approximate EEG recording from a subject during Stage 1 sleep. Draw two additional lines indicating approximately what the EEG recordings would look like when the subject is in Stage 4 sleep, and when the subject is awake and alert.

Stage 1:

Stage 4:

Awake and alert:

4. Connect the electrodes to the appropriate location on the subject below.

5. Draw arrows to the approximate location of the following structures associated with sleep and wakefulness: Hypothalamus, Reticular activating system, Medial pontine reticular formation.

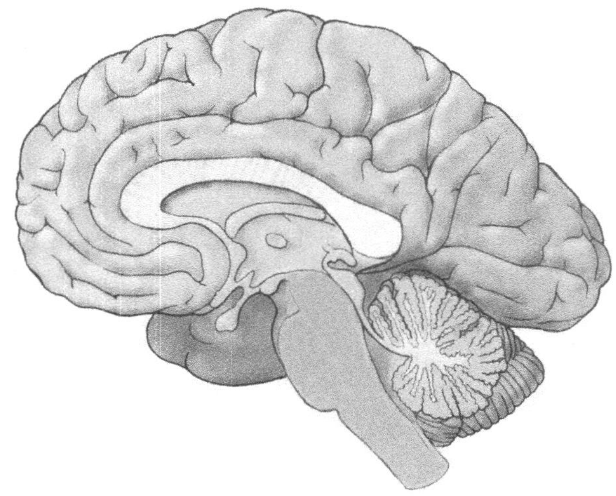

The Web

Consider using the following Web sites for additional information on some of the topics from this chapter:

1. National Sleep Foundation: www.sleepfoundation.org/
2. Restless Legs Syndrome Foundation: www.rls.org/
3. American Sleep Apnea Association: www.sleepapnea.org
4. Sleep disorders information page: www.sleepnet.com
5. Night Terror Resource Center: www.nightterrors.org/

CROSSWORD PUZZLE

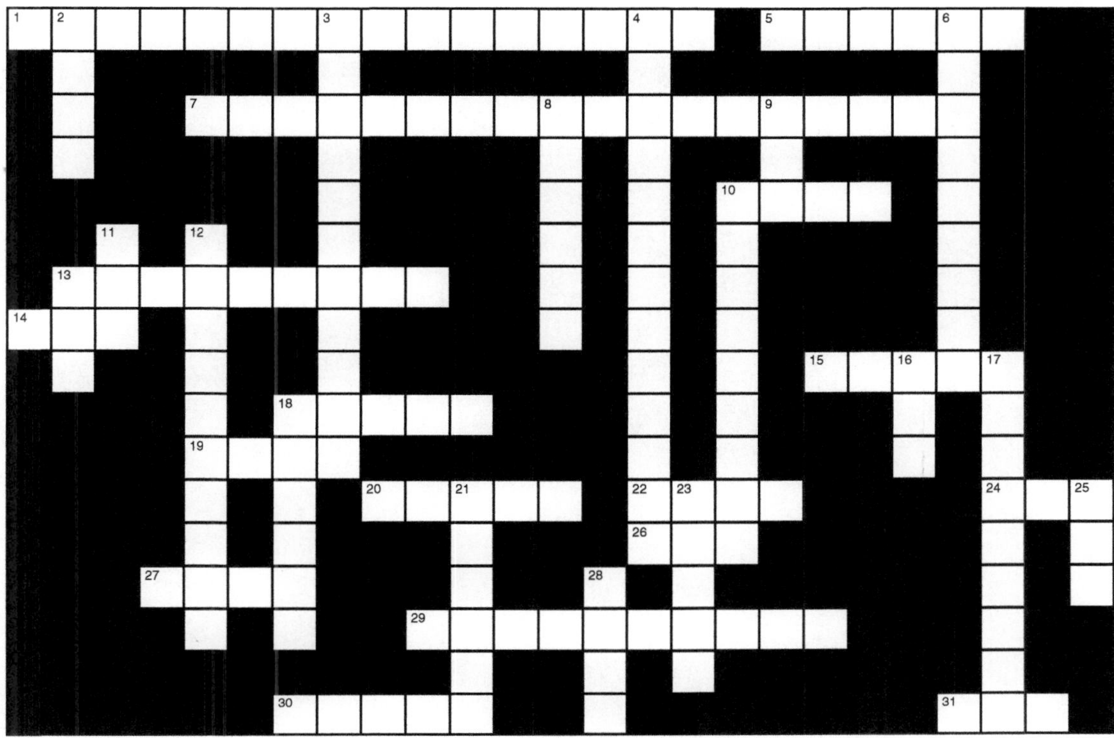

Across

1. Recordings of muscles, sometimes used in sleep reserch
5. Without 9 down, you might be running around the bedroom at night
7. Recordings of eyes, sometimes used in sleep research
10. 1 across; abbr.
13. _____ activating system is responsible for aspects of wake-sleep cycle
14. See 31 across
15. Latin word meaning "around"
18. Causes a type of insomnia; 18 across singular
19. Common term for 3 down; cat _____
20. Waves seen on 11 down when you begin to relax and enter sleep
22. Term for sleep during which dreams seldom occur
24. These really move during dreams
26. Disorder when people become depressed in the winter; abbr.
27. _____plexy can have you on the floor with laughter
29. There's another one in the pineal gland, see 8 down
30. Author of Interpretation of Dreams
31. With 14 across, term for disruptive time-zone travel effect

Down

2. When these become too restless during sleep, it is considered a sleep disorder syndrome
3. Technical term for a very short sleep session
4. Types include hypnogogic or LSD-induced
6. Person who unwillingly pulls all-nighters
8. One in the suprachiasmatic nucleus, another in the retina, with 29 across
9. Dream sleep; abbr.
10. When 8 down is reset by a zeitgeber, it is said to be _____
11. Scalp recordings of electrical activity during sleep; abbr.
12. Occurring at cycles of around a year
13. 13 across; abbr.
16. Proper term for 2 down sleep disorder; abbr.
17. Middle word in 26 across abbr., often associated with emotion
18. Disorders in which people have difficulty breathing during sleep
21. Term for the actual amount of time required to complete rhythm cycle
23. It's the way 24 across moves during dreams
25. 7 across singular; abbr.
28. State lacking consciousness, usually after severe brain damage

CHAPTER

14 How Do We Learn and Remember?

CHAPTER SUMMARY

In this chapter the concept of neuroplasticity is emphasized as a physiological basis for learning and memory. *Learning* is defined as a process that results in a relatively permanent change in behavior, whereas *memory* is the ability to recall or recognize previous experiences. Laboratory studies of learning and memory using animals have incorporated hundreds of tasks for animals. Among the earliest studies were those conducted by *Ivan Pavlov* using *classical conditioning* to teach associations between two stimuli. More recent variations of this type of conditioning are *eye-blink conditioning* and *fear conditioning*, both of which pair a conditioned stimulus (CS) to an unconditioned response (UCR) to produce a learned conditioned response (CR). These experiments have shown that separate pathways in the cerebellum and the amygdala are associated with eye-blink response and fear response, respectively. In this regard even simple forms of learning may be mediated by different brain regions or systems. *B. F. Skinner* expanded on earlier research conducted by *Edward Thorndike* to develop a learning paradigm now known as *instrumental conditioning*. Instrumental conditioning requires an animal to learn an association between its actions and the consequences of those actions.

Applying information from animal research to humans is difficult in part because much of human learning and memory is verbal. One general division of human memory seems to be the categories of *implicit* and *explicit memory*. In general terms, implicit memory refers to an unconscious memory (one that is not verbalized), whereas explicit memory refers to a conscious verbal memory. Implicit learning, also known as *procedural memory*, includes conditioned motor responses such as those involved in learning to ride a bicycle. Explicit memory, also known as *declarative memory*, includes recall of content of experiences such as remembering when you last rode a bicycle. The way in which these two types of memories are processed and stored seems to differ as well. Procedural memory utilizes a "bottom-up" processing scheme beginning, for example, with lower-order ("bottom") sensory feedback and muscular movements that are then related to, or incorporated into, higher-order cortical memories. Declarative memory utilizes a "top-down" processing scheme beginning with a higher-order ("top") cortical process of memory recall.

Distinctions can also be made between short-term and long-term memory. Short-term memory is held only for several minutes at most, and is controlled primarily by the frontal lobe. Long-term memory may hold information for a lifetime. Long-term verbal memories are controlled primarily by the temporal lobe. It should be noted, however, that even though some locations are more prevalent in learning and memory, it appears that nearly the entire nervous system can show plasticity responses with experiences.

In the 1920s Karl Lashley began a multidecade search for the location of a neural circuitry responsible for memory. His conclusion was that when damage occurs, the size of the area of damage was a greater factor in memory loss than was the location. Soon after Lashley's conclusion, a patient (H.M.) recovering from neurosurgery to reduce seizures exhibited near total loss of declarative memory as a result of the surgery. It was determined that damage to the anterior hippocampus, amygdala, and surrounding cortical structures were the cause of this memory loss. Since Lashley had produced primarily lesions of cortical regions and tested animals in procedural tasks, it was determined that his conclusion that cortical lesions had little if any effect on implicit memory was valid. Later research showed that loss of implicit memory (and subsequent disruption of procedural tasks) is produced by lesions of the basal ganglia, an area that Lashley had not examined in his lesion studies.

From the vast amount of data collected a theory of regions associated with explicit and implicit memory has begun to emerge. Explicit memory is currently believed to involve structures of medial temporal cortical regions, the frontal cortex, and closely related structures. *Alzheimer's disease* results in severe atrophy of tissue near the medial temporal region and is characterized by severe explicit-memory deficits. Similarly, *Korsakoff's syndrome* results in atrophy of the frontal lobe of most patients and is characterized by impairments of short-term explicit memory. The hippocampus likely plays an important role in visual and spatial contributions to explicit memories. Animals with hippocampal lesions are severely debilitated in visual-recognition tasks. In the same regard, animals that are particularly adept at hiding and then finding food stores tend to have large, well-developed hippocampal regions. Implicit memories appear to arise from basal-ganglia function stimulated by cortical inputs. The unconscious nature of these memories is accounted for by the lack of connections from the basal ganglia back to the cortex. *Parkinson's disease* is marked by degeneration of the basal ganglia and is characterized by deficits in implicit memory.

Emotional memory represents a unique category of memory. This type of memory is controlled primarily by the amygdala, and when dysfunctional may result in a pathology known as panic disorder. Lesions of the amygdala abolish emotional memory with little effect on implicit or explicit memory.

Santiago Ramón y Cajal first suggested in the 1920s that learning was likely a function of structural changes to the neuron. However, evidence substantiating this theory has only recently been produced. Researchers now know that the shape of dendrites show a great deal of plasticity, changing morphology in response to changing experiences of the organism. Gain or loss of synapses in turn results in changes in local circuitry. Very recent research also suggests that the central nervous system is also capable of producing new neurons in response to new experiences. For example, mice housed in an enriched environment (one containing many environmental stimuli) had more hippocampal neurons than mice raised in an impoverished environment. In humans, complexity of dendritic arbors is correlated with extent of formal education. Females have more complex dendritic arbors in Wernicke's area than males, correlating with superior verbal skills.

Several factors are capable of influencing neural plasticity. For example, the number of dendritic spines is positively correlated with the circulating estrogen seen in female rats during their four-day estrous cycle. *Glucocorticoids* produced by the adrenal gland in response to stress can be neurotoxic if maintained at a high level over an extended period of time, ultimately reducing the number of neurons, especially in the hippocampus. In addition to hormones, *neurotrophic factors* may also influence neural plasticity. One of these chemicals, *nerve growth factor*, is known to stimulate dendritic growth during development. *Brain-derived neurotrophic factor* (*BDNF*) is released during maze solving in rats and may act in a manner similar to nerve-growth factor in adult animals. Psychoactive drugs may also produce long-term changes in structure and function of synapses. Amphetamine and cocaine can increase dendritic growth and spine density in prefrontal-cortex and nucleus-accumbens neurons, resulting in sensitization of those synapses to future administration of the drug.

Neuronal plasticity is perhaps most apparent when assessing recovery from brain damage. The "three-legged cat solution" suggests that when behavioral function is lost as a result of injury, behavioral modifications can be made to compensate (as might be seen in a cat learning to walk on three legs). Such behaviors are undoubtedly mediated by synaptic changes associated with learning new skills. The new circuit solution suggests that following neural damage, new synaptic connections may form or old synapses may become more active in an effort to mediate more behaviors with less neural tissue. This type of plasticity may be enhanced by pharmacological intervention, such as the administration of nerve growth factor shortly after the damage has occurred. There is now evidence that neuronal transplants may also be useful for reversing some deficits following brain damage. However, this procedure is currently limited to only a few types of damage, such as Parkinson's disease. Finally, research with epidermal growth factor (EGF) has shown that cells lining the ventricles may be stimulated to reproduce and migrate to some regions of the brain. This research is very recent and encouraging; however, researchers have yet to determine how to encourage newly migrated cells to establish functional connections as the next step in reversing deficits.

KEY TERMS

The following is a list of important terms introduced in Chapter 14. Give the definition of each term in the space provided.

Learning and Memory

Neuroplasticity

Learning

Memory

Pavlovian conditioning

Respondent conditioning

Classical conditioning

Eye-blink conditioning

Fear conditioning

Instrumental conditioning

Operant conditioning

Visuospatial learning

Learning set

Implicit memory

Explicit memory

Declarative memory

Procedural memory

Episodic memory

Neural Systems: Implicit/Explicit

Alzheimer's disease

Korsakoff's syndrome

Anterograde amnesia

Retrograde amnesia

Emotional memory

Structural Basis of Plasticity

Brain-derived neurotrophic factor (BDNF)

Enrichment

Glucocorticoids

Neurotrophic factors

Nerve growth factor (NGF)

Drug-induced behavior sensitization

Recovery from Brain Injury

Three-legged cat solution

Epidermal growth factor

New-circuit solution

Lost-neuron-replacement solution

KEY NAMES

The following is a list of important names introduced in Chapter 14. Explain the importance of each person in the space provided.

Ivan Pavlov

Edward Thorndike

B. F. Skinner

Karl Lashley

Santiago Ramón y Cajal

PRACTICE TEST

Multiple-Choice Questions

Answer each of the following multiple-choice questions with the best possible answer based on information from your text.

1. Which of the following terms refers to the type of learning studied by Ivan Pavlov?
   A. Pavlovian conditioning
   B. respondent conditioning
   C. classical conditioning
   D. All of the answers are correct.
   E. None of the answers is correct.

2. When fear-conditioning a rat evokes an anxiety response from a tone that has been paired to an electrical shock, the tone represents which of the following?
   A. conditioned response
   B. conditioned stimulus
   C. unconditioned response
   D. unconditioned stimulus
   E. could represent any of the answers depending on design of the study

3. Edward Thorndike was one of the earliest researchers to examine instrumental conditioning. However, ________________ is generally the researcher associated with this type of conditioning.
   A. Ivan Pavlov
   B. Richard Morris
   C. B. F. Skinner
   D. Karl Lashley
   E. Santiago Ramón y Cajal

4. An amnesic person is taught a new motor task. Three days later that person shows a high level of retention for the motor functions, but cannot recall ever learning the task. This person is showing which of the following?
   A. explicit memory with no implicit memory
   B. implicit memory with no explicit memory
   C. both explicit memory and implicit memory
   D. neither implicit nor explicit memory
   E. amnesic patients cannot be tested for memory recall

5. The term *top-down* processing is sometimes used to describe the process used in declarative memory. From the term *top-down* processing, what does the word *top* refer to?
   A. dendrite
   B. axon terminal
   C. brainstem
   D. cortex
   E. pituitary gland

6. Memory is often divided into short-term and long-term memory. Long-term verbal memories of verbal information appear to be stored in the temporal region. What region appears to play an important role in temporary, short-term memories?
   A. frontal
   B. temporal
   C. parietal
   D. occipital
   E. Short-term memories are not influenced by cortical regions.

7. After bilateral removal of hippocampal/amygdala regions, H.M. showed severe impairment in which of the following?
   A. perceptual tests
   B. intelligence
   C. recall of childhood memories
   D. explicit memory
   E. All of the answers are correct.

8. At the age of 78, patient J.K. began to show significant deficits in implicit memory that were believed to result from degeneration of the basal ganglia. Which of the following was responsible for the basal-ganglia degeneration, and ultimately the memory impairments, seen in J.K.?
   A. Korsakoff's syndrome
   B. Alzheimer's disease
   C. Parkinson's disease
   D. severe seizures
   E. normal aging

9. Which of the following has been determined to be the cause of Alzheimer's disease?
   A. genetics
   B. exposure to toxins in adulthood
   C. exposure to toxins during fetal development
   D. abnormal rapid aging of neurons
   E. The cause of Alzheimer's disease is unknown.

10. Korsakoff's syndrome can result in severe memory deficits. Which of the following is the cause of Korsakoff's syndrome?
    A. alcohol consumption
    B. thiamine deficiency
    C. aging
    D. genetics
    E. head trauma

11. Which of the following neurotransmitters is released in high concentration in the basal ganglia, and as such implicated in implicit memory?
    A. serotonin
    B. acetylcholine
    C. dopamine
    D. norepinephrine
    E. epinephrine

12. What effect would you expect in an animal that had bilateral lesions of the amygdala?
    A. loss of implicit memory
    B. loss of explicit memory
    C. loss of emotional memory
    D. loss of short-term memory
    E. All of the answers are correct.

13. Morphological changes in which of the following most likely represent(s) neural changes associated with learning and memory?
    A. glial cells
    B. myelin sheaths
    C. the blood–brain barrier
    D. dendritic arbors
    E. nuclear DNA

14. Elizabeth Gould and her colleagues recently provided evidence that new neurons may form in response to learning explicit-memory tasks. In what structure did these researchers report genesis of these neurons?
    A. frontal cortex
    B. amygdala
    C. hippocampus
    D. olfactory bulb
    E. temporal lobe

15. Researchers have long speculated that morphological changes in the brain could be correlated to learning. When Scheibel and colleagues examined Wernicke's area tissue from deceased patients, which of the following did they find?
    A. more neurons in college-educated patients than in high school–educated patients
    B. more glial cells in college-educated patients than in high school–educated patients
    C. more dendritic arbors in college-educated patients than in high school–educated patients
    D. All of the answers are correct.
    E. None of the answers is correct.

16. Which of the following hormones has been shown to affect neural structure in adult rats?
    A. estrogen
    B. growth hormone
    C. insulin
    D. All of the answers are correct.
    E. None of the answers is correct.

17. In animals, brain-derived neurotrophic factor (BDNF) is released and thought to promote structural changes in neurons during:
    A. maze learning.
    B. REM sleep.
    C. NREM sleep.
    D. fetal development.
    E. sexual behavior.

18. Robinson and Kolb have shown that when rats are sensitized to amphetamine through repeated injections, there is a subsequent:
    A. decrease in release of dopamine.
    B. loss of glial cells in dopamine-rich regions.
    C. loss of neurons in dopamine-rich regions.
    D. increase in neurons in dopamine-rich regions.
    E. increase in growth of dendrites and spine density in dopamine-rich regions.

19. The therapy of replacing lost neurons to promote recovery of lost behavioral function in humans with degenerative diseases has been most successful when using cells from which of the following?
    A. rats
    B. pigs
    C. monkeys
    D. human fetuses
    E. None of the answers is correct.

20. Epidermal growth factor (EGF) has been most successful at promoting neural growth when injected into what region?
    A. basal ganglia
    B. hippocampus
    C. frontal cortex
    D. amygdala
    E. ventricles

Short-Answer Questions

Answer each of the following questions with a brief but complete written answer based on information from your text.

1. Describe in basic terms the difference between classical and operant conditioning.

2. Describe in general terms the difference between implicit and explicit memory.

3. Briefly describe the characteristics of episodic amnesia.

4. For a large part of his career, Karl Lashley searched for a neural circuit for memory, using knife-cut lesions of the cortex in rats in an attempt to disrupt their maze-learning ability. Give at least two reasons why Lashley was unsuccessful in his attempts to disrupt memory using these methods.

5. Briefly describe the type of surgery that was conducted on patient H.M. and the result of that surgery. Include in your description three brain structures affected by the surgery.

6. In what type of animal would you expect to find a large, well-developed hippocampus? Explain your answer.

7. Briefly describe the cause of Korsakoff's syndrome, the associated memory impairment, and the affected brain region most likely responsible for the impairment.

8. In an effort to show neuroplasticity in adults, researchers have assessed brains of individuals with highly skilled motor functions (such as playing a string instrument), and brains of animals that have undergone amputation of a limb. Briefly describe how the brain responds in each of these cases and what these findings tell us about the capacity for neural reorganization in the adult brain.

9. Briefly explain what is meant by the "three-legged cat solution" when describing recovery from neural damage.

10. According to the new circuit solution of recovery from neural damage, what two forms of therapy would you suggest for a patient? Explain why such therapies are thought to be beneficial.

Matching Questions

Complete each of the following matching questions based on information from your text.

1. Identify each type of learning as operant conditioning (OC) or classical conditioning (CC).

   A. ___ A dog waiting near the table to be fed food scraps
   B. ___ Ducking your head when you hear a loud noise
   C. ___ The feeling of hunger when you smell pizza
   D. ___ Studying late into the night before an exam
   E. ___ Holding the door open for someone entering a building behind you

2. Identify each of the following as more closely associated with implicit memory (IM) or explicit memory (EM).

   A. ___ Riding a bicycle
   B. ___ Childhood memories
   C. ___ Declarative memory
   D. ___ Procedural memory
   E. ___ Top-down processing
   F. ___ Bottom-up processing

3. Match each of the following neural structures or regions to the memory with which they are most closely associated.

A. Hippocampus	___ Visual object memory
B. Frontal lobe	___ Short-term memory
C. Amygdala	___ Emotional memory
D. Basal ganglia	___ Implicit memory
E. Perirhinal cortex	___ Object location

4. Match each of the following disorders with the appropriate symptom or characteristic.

	___ Degeneration of frontal lobe
A. Parkinson's disease	___ Loss of implicit memory
B. Alzheimer's disease	___ Degeneration of entorhinal cortex
C. Korsakoff's syndrome	___ Caused by a vitamin deficiency
	___ Degeneration of the basal ganglia

5. Match each of the following mechanisms for promoting recovery of behavioral function with the appropriate description or feature.

	___ Recommends use of speech therapy
A. Three-legged cat solution	___ Undamaged regions will assume control of behavior
B. New circuit solution	___ Recommends use of nerve growth factor
C. Replacing lost neurons	___ Recommends use of epidermal growth factor (EGF)
	___ Has been used in many Parkinson's patients

Diagrams

1. On the diagram of the brain below, identify the approximate location of the following regions: superior temporal gyrus, inferior temporal gyrus, middle temporal gyrus.

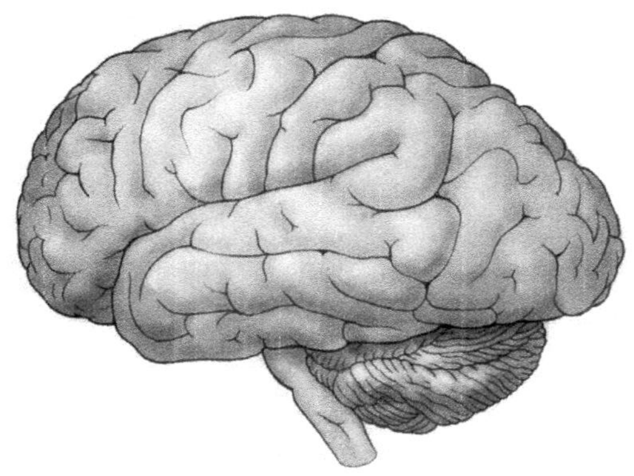

2. On the diagram of the ventral surface of the brain below, identify the approximate location of the following structures: amygdala, hippocampus, entorhinal cortex, parahippocampal cortex, perirhinal cortex.

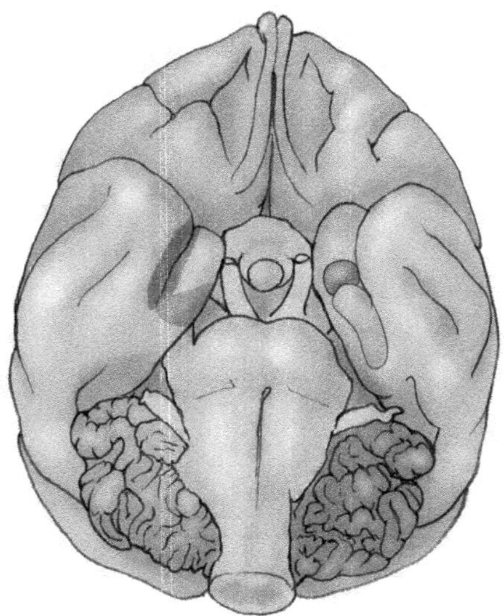

3. Complete the schematic below by indicating the appropriate structure letter in the correct box to show the circuit proposed for implicit memory.
   A. Rest of cortex
   B. Sensory and motor information
   C. Substantia nigra–dopamine
   D. Basal ganglia
   E. Ventral thalamus
   F. Premotor cortex

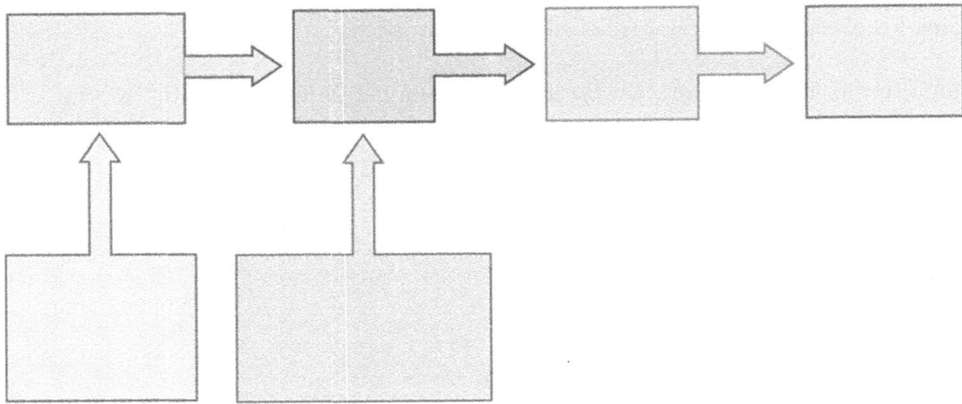

4. Below is a depiction of a dendrite with dendritic spines during times of moderate estrogen concentration. Indicate how you would expect dendrites to change with varying levels of estrogen concentration.

5. Rats exposed to repeated administration of amphetamine exhibit significant structural changes in some neurons. Draw a neuron depicting changes that might be seen in such a "sensitized" rat, as compared to the diagram below depicting a normal neuron.

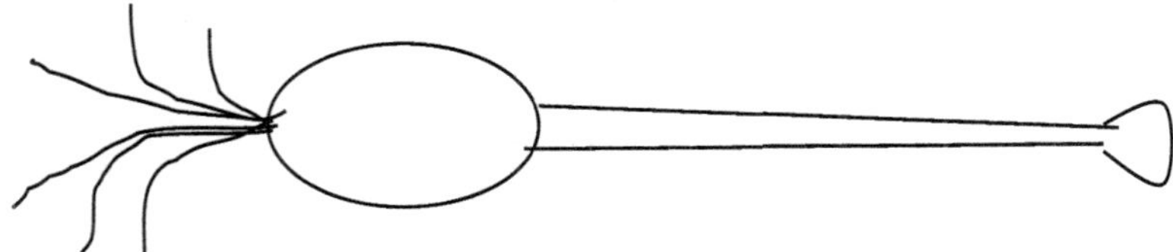

The Web

Consider using the following Web sites for additional information on some of the topics from this chapter:

1. Alzheimer's Association: www.alz.org/
2. Amnesia resource page: www.memorylossonline.com/resources.htm
3. Coma Recovery Association, Inc.: www.comarecovery.org
4. Hippocampus information: www.psycheducation.org/emotion/hippocampus.htm
5. The B. F. Skinner Web page: www.bfskinner.org/

CROSSWORD PUZZLE

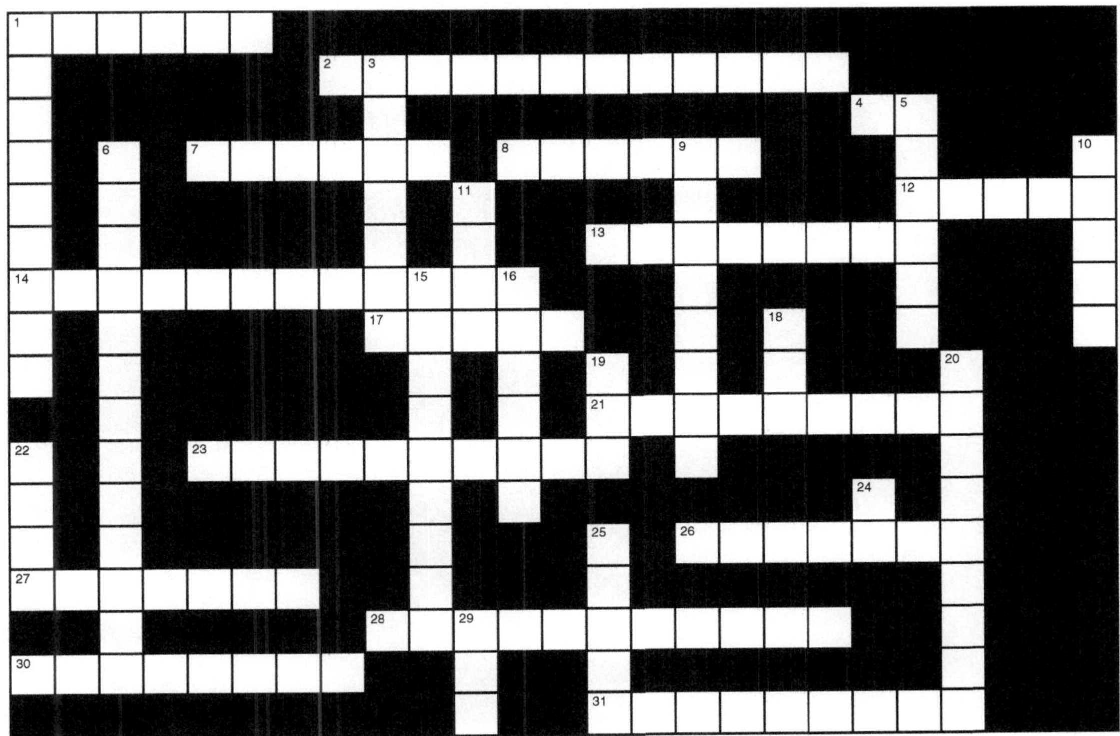

Across

1. The fellow who pioneered 20 down/1 down/23 across, Ivan to his friends
2. 14, 23 across, 1, 3, 20 down are all followed by this word
4. 27 across, to his friends
7. Garcia kept these from eating sheep with taste aversion learning
8. Primate used for some learning research
12. Perhaps the first to suggest learning was associated with morphological changes, Ramón to his friends
13. Basal ganglia memory, like how to ride a bike
14. Another term for 3 down
17. Number of limbs 29 down had to work with after amputation
21. The "E" of EGF
23. Least common of the three terms used for conditioning type in 20 down and 1 down
26. NGF is one of many trophic _____
27. He followed the work of 15 down, using pigeons and rats mostly
28. Memory that remembers the process of learning
30. Korsakoff's technically is not a disease but rather a _____
31. A type of learning associated with stimuli that evokes a lot of anxiety, for example

Down

1. Another term for 20 down, this one named for Ivan
3. 27 across studied learning using this type
5. The "F" in NGF that promotes recovery of neural function
6. Increased behavioral response after experiencing a stimulus, opposite of habituation
9. Memory that's not implicit
10. With 18 down, a type of conditioning (learning) using a puff of air
11. A type of conditioning based on extreme anxiety, usually with electrical floor grid
15. One of the first to study 3 down, Ed to his friends, he used cats
16. Recovery solution named for an animal amputee with 29 down and 17 across
18. See 10 down
19. Understanding a solution may be used for many circumstances is a learning ___
20. Term for S-R conditioning or Beethoven's music genre
22. Drinking enough of these filled with beer over the years could lead to Korsakoff's
24. Boxing results that could lead to memory loss; abbr.
25. The "N" in NGF
29. Animal for which recovery solution in 16 down was named

CHAPTER

15 How Does the Brain Think?

CHAPTER SUMMARY

In trying to determine how the brain "thinks," it appears that many systems are likely involved. In the mammalian brain, these systems are thought to be located primarily in the neocortex. Because "thought," "emotion," "motivation," and other aspects of the human experience have no physical basis, they must be studied as *psychological constructs*. Psychological constructs, in simple terms, are descriptions of processes based on information or impressions gathered from observations. Psychologists often use the term *cognition* to describe thought processes. Cognition is the ability to attend to internal or external stimuli and to respond to these stimuli in a meaningful manner. Human thought is unique, in that it is closely tied to language. Although other animals have the cognitive ability to interpret vocalizations, they lack the ability to structure such vocalization into *syntax*. Syntax thus represents a particularly important element in the development of complex human thought. Syntactic language is believed to have evolved from a tendency of the human brain to order strings of events, movements, or thoughts. A lack of this general tendency to order output in nonhuman animals suggests that this may be the most critical feature in the development of complex thought. Furthermore, ordering behavior seems to be controlled by the neocortex in the frontal lobe, one of the most highly developed regions in the human brain compared to other animals. Thus, although many animals are likely capable of thought, the human brain has developed the most complex abilities for this feature.

It is difficult to determine the neural basis for thought, since it likely involves a variety of interconnected units depending on the output. However, it has been shown that even a single neuron is capable of responding to a particular stimulus. Thus, the basis for cognition is likely a convergence of many such neurons linked together as cohesive units. The process of cognition likely results as information is collected from many of these basic units of neural processing and a single higher-level structure produces a response. One of the highest-level regions of brain functioning is the *association cortex*. The association cortex receives primary input from the thalamus and other cortical regions that have already undergone a great deal of convergence and processing. The association cortex is believed to then process this higher-level information with consideration to stored knowledge about the internal and external environment. For example, visual input of an object is processed to determine what the object is and where the object is located using the association cortex in the temporal and parietal lobes, respectively.

There appears to be a correlation between fine motor control and some cognitive processes, including spatial cognition. It is believed that discrete movements are interpreted and integrated, particularly in the right parietal association cortex, to create spatial mapping. Attending to subtle cues also appears to be essential to complex cognitive function. The association cortex in the frontal and parietal lobes is particularly involved in attention. Damage to the parietal lobe of one hemisphere may result in *neglect* of one half of the body and visual field, including not dressing or grooming one half of the body and not being conscious of objects in one half of the field of vision.

Related to processes of ordering and attending to stimuli, the frontal lobe is responsible for planning behaviors. Individuals with frontal-lobe damage exhibit a great deal of difficulty organizing and planning a series of behaviors to be executed in the future (known as temporal organization). One sensitive test of frontal-lobe damage is the Wisconsin Card Sorting Task that requires a series of shifts from one sorting strategy to a new strategy.

It was shown that neurons in the frontal lobe of monkeys also have a capacity to "understand" movements of other monkeys. These cells (called *mirror cells*) became active when the subject performed a particular movement, but more interesting, these same cells became active when the subject viewed another monkey performing the same movement. These findings suggest mirror cells in the frontal association cortex are capable of recognizing external stimuli in the context of internal stimuli. In simple terms, mirror cells "understand" some aspect of the external stimuli. PET scans suggest mirror cells exist in humans as well, but are localized primarily to the left hemisphere of the frontal lobe.

The study of the human brain and behavioral function, known as *cognitive neuroscience*, was long limited to assessing human behavior following damage produced by strokes, tumors, disease, and so on. Technological advances have recently begun to help us increase our understanding of brain functions associated with thought processes without reliance on brain damage. Such technology includes PET scans, as previously mentioned, along with *magnetic resonance imaging* (*MRI*) and *functional MRI* (*fMRI*). MRI uses magnetic fields to produce a relatively high resolution image of the brain. fMRI uses this same method with the addition of locating regions of higher oxygen consumption to identify regions that become particularly active during mental tasks.

The fact that some processes of human thought appear to be dominated by a particular hemisphere is referred to as *functional asymmetry*. Functional asymmetry likely results in part from *anatomical asymmetry*, the tendency for structures to be larger in one hemisphere. One example of this is the larger volume of sensorimotor cortex representing the face in the left hemisphere of the brain. Presumably this anatomical asymmetry is directly related to the functional asymmetry of language production in this hemisphere. Analyses of functional and anatomical asymmetry have suggested that right parietal regions of association cortex control thought associated with spatial cognition and music appreciation, while the same region of the left hemisphere controls thought associated with language, reading, and arithmetic. Some individuals have had their corpus collosum cut in an effort to reduce epileptic seizures. The result of this surgery (in addition to reduction of seizures) is that the two hemispheres are unable to communicate, providing a unique opportunity for studying functional asymmetry. One interesting finding from these patients is that when an object is presented only in the left visual field (to the right hemisphere), the subject reports not seeing any object. However, when presented with the opportunity to select the object from a group of objects with the left hand (controlled by the right hemisphere), the subject will choose the correct object. It appears that the left hemisphere, having neither seen the object nor having received input from the right hemisphere about the object, responds verbally that it has seen nothing. The right hemisphere, though aware of the object, is incapable of producing language and thus does not respond to the query of object identification. Also interesting is the finding that although the left hemisphere appears to control comprehension of language, comprehension of Braille reading is controlled by the right hemisphere. This finding is likely due to the fact that Braille requires spatial patterning for comprehension.

In addition to hemispheric differences, there are also gender differences in cognitive organization. In general terms, females show greater proficiency in verbal tasks, whereas males show greater proficiency in spatial tasks. Differences appear to be mediated by gonadal hormones, since they are not apparent in rats that have had gonads removed at birth. In human females, the increase in verbal fluency appears to coincide with the onset of puberty, suggesting an organizational effect of ovarian hormones. In addition, language disruption is more likely to occur with damage to the posterior cortex of males and the anterior cortex of females. One theory for gender differences in humans is that evolution may have favored males with spatial abilities that enhanced hunting and gathering strategies. Females, on the other hand, would benefit more from developing fine motor and communication skills that would enhance rearing offspring.

The relationship of handedness to hemispheric control of language is significant but not absolute. Virtually all right-handed individuals have left-hemisphere language centers. Approximately 70 percent of left-handed individuals also have left-hemisphere language centers. The remaining left-handed individuals are about evenly split between having right-hemisphere language centers and bilateral language centers (also called *anomalous speech representation*). The neurological reason for handedness, not being a function of hemispheric dominance, remains unknown.

Intelligence is often referred to, but is ill defined. Using all measures of intelligence to date, it has been determined that brain size correlates poorly to intelligence. Case studies of Einstein's brain have revealed some interesting findings of specific regions being larger, but these data are far from conclusive. Some researchers believe there are numerous categories of intelligence, including not only mathematical and verbal skills, but also navigation and social skills, among others. Most researchers agree that intelligence may be measured in terms of *convergent thinking* and *divergent thinking*. In general terms, convergent thinking refers to the ability to generate a single answer to a specific question (such as mathematical solutions), whereas divergent thinking requires generating numerous possible answers to a single question (such as determining all the possible ways to get a high score on your final exam). In terms of these types of intelligence, damage to the parietal lobes results in disruption of convergent thought. Divergent thought processes, on the other hand, are selectively disrupted with damage to the frontal lobes. Regardless of how intelligence is categorized or subdivided, there appear to be both genetic and environmental contributions to the neural basis for intellectual thought.

The concept of consciousness, or awareness of our mental capacities, is likely a collection of processes including sensory perception, thought, emotion, language, and so on. Research suggests that some perception occurs at a level below consciousness. Some neurological disorders also emphasize this point. Previous chapters have described blindsight, agnosias, and implicit learning in amnesia. These disorders illustrate that perception and even learning can occur without conscious knowledge. On the other hand, it appears consciousness may be stimulated without appropriate perception. This is a more difficult concept, but consider visual hallucinations as an example. In this case, there is conscious perception of a nonexistent stimulus. In other words, the nervous system is responsible for both unconscious and conscious thought, and these processes (though intimately linked) are controlled separately.

KEY TERMS

The following is a list of important terms introduced in Chapter 15. Give the definition of each term in the space provided.

Characteristics of Thinking

Psychological constructs

Cognition

Syntax

Cognition and the Association Cortex

Spatial cognition

Attention

Neglect

Extinction

Planning

Studying Brain and Cognition

Cognitive neuroscience

Magnetic resonance imaging (MRI)

Functional MRI (fMRI)

Diffusion tensor imaging (DTI)

Cognitive neuroscience

Cerebral Asymmetry

Apraxia

Split brain

Corpus callosum

Cognitive Organization

Anomalous speech representation

Synesthesia

Intelligence

Convergent thinking

Divergent thinking

Intelligence A

Intelligence B

Consciousness

PRACTICE TEST

Multiple-Choice Questions

Answer each of the following multiple-choice questions with the best possible answer based on information from your text.

1. Which of the following best describes a psychological construct as defined in your text?
   A. an element of genetic coding required for thought
   B. an element of a neuron required for thought
   C. a system of neurons within a brain structure required for thought
   D. a brain structure required for thought
   E. a mental process that cannot be seen but is inferred to exist as a basis for thought

2. More important for human thought than words is the process of being able to string together words into grammar. What is another term for this process?
   A. ambulating
   B. syntax
   C. cognition
   D. constructing
   E. language

3. The process of stringing together words to create language is likely one of the most prevalent features of advanced thought in humans. Language, however, is not the only example of our tendency to combine behaviors into complex outputs. Which of the following represents another example of this ability?
   A. eating a meal
   B. sleeping at night
   C. choreographing dances
   D. engaging in sexual activities
   E. All of the answers are correct.

4. Information sent to the association cortex may be thought of as which of the following?
   A. direct from sensory systems
   B. direct from motor systems
   C. highly fragmented
   D. highly processed by other brain structures
   E. having little consequence on thought

5. Attention is essential for proper cognitive functioning. What happens in individuals who suffer damage to the parietal association cortex in one hemisphere that disrupts attention?
   A. They may exhibit unilateral neglect.
   B. They may fail to dress one side of their body.
   C. They may not see objects presented in half of their visual field.
   D. They may fail to respond to commands to move the limbs on one side of the body.
   E. All of the answers are correct.

6. The Wisconsin Card Sorting Task is an excellent screening tool for damage to which of the following regions?
   A. parietal lobe damage
   B. frontal lobe damage
   C. spinal damage
   D. visual inattention
   E. hippocampal damage

7. Although mirror neurons have been found in both monkeys and humans, one fundamental difference between the two is that these neurons are:
   A. larger in humans.
   B. smaller in humans.
   C. localized to the left hemisphere in humans.
   D. localized to the right hemisphere in humans.
   E. found in all lobes of the brain in humans.

8. MRI has become a very useful research tool for those studying cognition in the human brain. What does MRI stand for?
   A. Metabolic Resistance Imaging
   B. Metabolic Resolution Imaging
   C. Metabolic Reaction Imaging
   D. Metabolic Resonance Imaging
   E. None of the answers is correct.

9. The most likely reason that the sensorimotor cortex representing the face in the left hemisphere is larger than the same region in the right hemisphere is that it is responsible for:
   A. language interpretation.
   B. language production.
   C. expression of emotions.
   D. spatial representation.
   E. All of the answers are correct.

10. Which of the following statements is true of hemispheric asymmetry?
    A. It is apparent only in humans.
    B. It is apparent only in humans and some higher-order primates.
    C. It is apparent only in humans and all other primates.
    D. It is apparent in most animals with the exception of birds.
    E. It is apparent in most animals, including many species of birds.

11. According to research results from the Kimura lab, which ear should you turn toward your professor during a lecture and which ear should you turn toward the jukebox at the bar?
    A. right ear toward professor, left ear toward jukebox
    B. left ear toward professor, right ear toward jukebox
    C. left ear toward both professor and jukebox
    D. right ear toward both professor and jukebox
    E. Kimura found no difference in ear preference for language and music.

12. Split-brain individuals provide a unique opportunity to assess independent functioning in the two hemispheres. What is the name of the structure that is severed to produce these individuals?
    A. thalamus
    B. hypothalamus
    C. basal ganglia
    D. frontal cortex
    E. None of the answers is correct.

13. Which of the following appears to be the most likely contributing factor to development of superior verbal fluency in females when compared to males?
    A. lack of testosterone during fetal development
    B. presence of estrogen during fetal development
    C. influence of estrogen at puberty
    D. social influences during childhood
    E. None of the answers is correct.

14. An injection of sodium amobarbital into the left carotid artery would have which of the following effects on most individuals?
    A. It would enhance speech ability.
    B. It would impair speech ability.
    C. It would enhance spatial ability.
    D. It would impair spatial ability.
    E. It would result in death.

15. A person with anomalous speech representation:
    A. has a speech center in the left hemisphere.
    B. has a speech center in the right hemisphere.
    C. is more likely to be left-handed than right-handed.
    D. All of the answers are correct.
    E. None of the answers is correct.

16. Synesthesia is a condition in which:
    A. An individual is not responsive to anesthetics.
    B. A child cannot learn to produce language.
    C. Sensory experiences cross sensory modalities.
    D. The two hemispheres of the brain develop with no anatomical difference.
    E. A male develops brain structures in a female pattern.

17. In the 1920s, Charles Spearman proposed that differences in brain architecture could result in the capacity for high or low "g" intelligence. What does the "g" stand for?
    A. general
    B. gonadal
    C. geographical
    D. group
    E. geometric

18. In classifying types of intelligence, J.P. Guilford refers to traditional tests of knowledge and reasoning that require a single narrowly defined answer, as tests of which of the following?
    A. general intelligence
    B. convergent thinking
    C. divergent thinking
    D. rational thinking
    E. insight

19. Donald Hebb proposed that all individuals have two forms of intelligence. Intelligence A is innate potential. Intelligence B is intelligence that may be influenced by which of the following?
    A. disease
    B. injury
    C. experience
    D. exposure to toxins
    E. All of the answers are correct.

20. Regarding consciousness, it is possible to perceive stimuli without being consciously aware of that perception. In this text, you have read about several disorders with symptoms that include lack of conscious awareness of stimuli. Which of the following is *not* an appropriate example of a disorder that results in this phenomenon?
    A. blindsight
    B. form agnosia
    C. implicit learning in amnesia
    D. visual neglect
    E. phantom-limb pain

Short-Answer Questions

Answer each of the following questions with a brief but complete written answer based on information from your text.

1. Briefly explain what the term *cognition* means and how cognition relates to thought.

2. Much of the information in this chapter suggests that stringing together sequences of behaviors is a high-level approach to thinking. Give a brief example of how successfully stringing together behavioral output could be advantageous to survival of an animal.

3. The frontal lobe is thought to be essential for temporal organization of behavior. Briefly explain what is meant by temporal organization, including a simple example.

4. Briefly explain when “mirror neurons” fire and the proposed function of these cells.

5. Describe the basic research methods used in the field of cognitive neuroscience.

6. Briefly explain what is meant by the terms *anatomical asymmetry* and *functional asymmetry*. Give a simple example of each.

7. If a split-brain person were exposed to a coffee mug in her left visual field and then asked to identify the object, she would reply that she had not seen an object. However, if allowed to select the object she had just seen from a group of objects, she would accurately choose the coffee mug. Briefly explain the neurological basis for this phenomenon.

8. Briefly describe the phenomenon of synesthesia.

9. Briefly explain what is meant by convergent thinking and divergent thinking. Give a brief example of each.

10. Donald Hebb proposed categorizing human intelligence into two basic forms. List the two categories he proposed and include a brief description of each.

Matching Questions

Complete each of the following matching questions based on information from your text.

1. Match the following brain regions with the most appropriate cognitive function.

	___ Map reading
A. Left hemisphere	___ Language production
B. Right hemisphere	___ Temporal planning
C. Frontal lobe	___ Music appreciation
	___ Language comprehension

2. Identify each of the following as being associated with convergent thinking (CT) or divergent thinking (DT).

___ Disrupted particularly by frontal-lobe damage
___ Measured with traditional intelligence tests
___ Solving arithmetic problems or defining words
___ Used to generate numerous answers for a single question
___ Disruption is often associated with apraxia or aphasia

3. Match the following neurological phenomena with the appropriate feature or description.

A. Split brain	___ Ignoring sensory information, usually on one side of the body
B. Apraxia	___ Inability to make voluntary movements
C. Frontal-lobe damage	___ Causes difficulty with the Wisconsin Card Sorting task
D. Synesthesia	___ Caused by damage to the corpus callosum
E. Neglect	___ Joining sensory experiences across sensory modalities

4. Match each of the following terms with the appropriate description or definition.

A. Construct	___ Pattern or structure of word order in a phrase
B. Syntax	___ The temporal organization of behavior
C. Spatial cognition	___ Idea resulting from a set of impressions
D. Extinction	___ A form of neglect
E. Planning	___ Term for a wide range of mental abilities

5. Identify each of the following as an object that could be identified and verbally described (IV) or an object that could not be verbally described (NV) by a split-brain patient with a left hemisphere language center.

___ Object placed in left hand
___ Object placed in right hand
___ Object briefly flashed in left visual field
___ Object briefly flashed in right visual field
___ Object briefly flashed in the center of the visual field

Diagrams

1. Label the following regions of the primary association cortex on the diagram below: Primary motor, Primary sensory, Primary visual, Primary auditory.

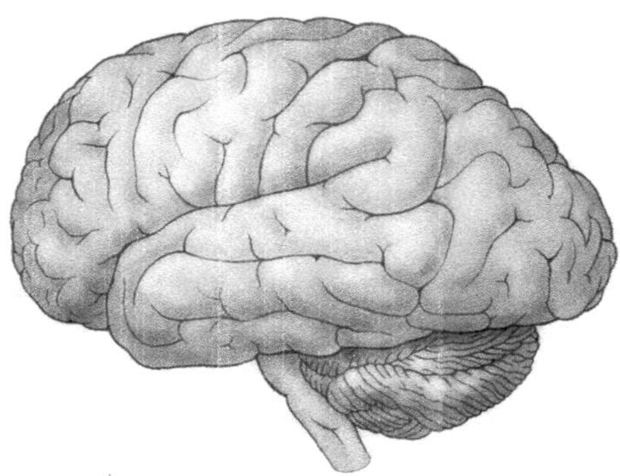

2. Below are diagrams of two individuals, each with a severed corpus callosum. Indicate the appropriate neural pathway taken by incoming visual information for each individual. Also indicate which individual will respond correctly when asked to name the object flashed on the screen in front of them.

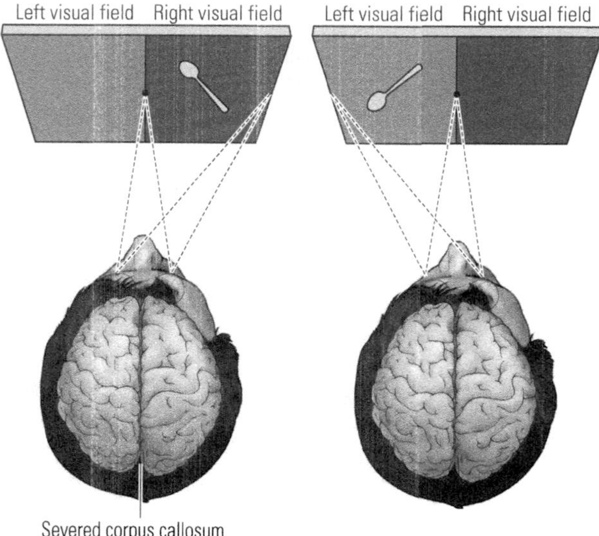

3. Which of the following best represents a comparison of neurons from the prefrontal region between male and female rats that had their gonads or ovaries removed at birth?

A.		B.		C.	
Male	Female	Male	Female	Male	Female

4. Which of the following would have the most difficulty with the task of drawing the accompanying figure?
   A. split-brain person using left hand
   B. split-brain person using right hand
   C. person with intact brain using right hand
   D. person with intact brain using left hand

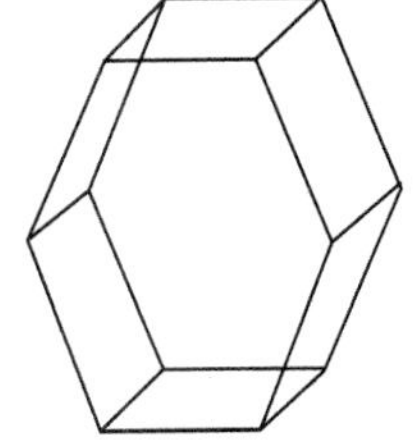

5. According to Doreen Kimura, males and females are almost equally likely to be aphasic following left-hemisphere lesions. On the diagram below, indicate the two primary regions where damage may produce aphasia. In addition, indicate which of the two regions, when damaged, is more likely to produce apraxia in males and which region is more likely to produce apraxia in females.

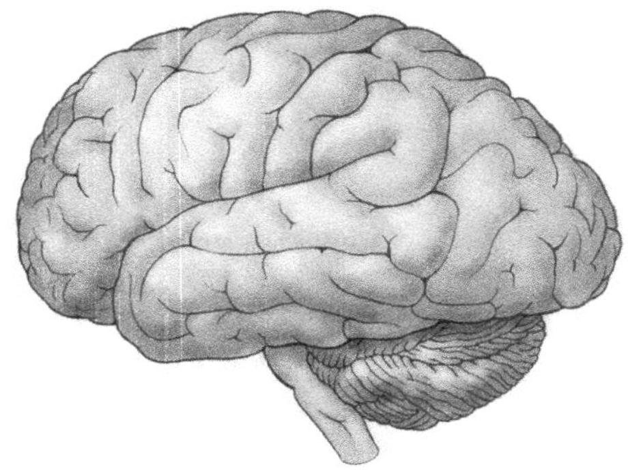

The Web

Consider using the following Web sites for additional information on some of the topics from this chapter:

1. Cognitive Neuroscience Society: www.cogneurosociety.org
2. Humorous and educational split brain site: www.nobelprize.org/educational/medicine/split-brain/
3. Synesthesia and the synesthesic experience: web.mit.edu/synesthesia/www/synesthesia.html
4. Functional neuroimaging: www.functionalmri.org
5. Apraxia Kids: www.apraxia-kids.org/

CROSSWORD PUZZLE

Across

1. Study of the neural basis of 12 across is often called cognitive _____
5. Word added to the beginning of 25 down when used for real-time research
6. Frontal association cortex is particularly necessary for this aspect of thought
7. Medial forebrain bundle; abbr.
9. Term for patterns and structure used for words in human language
10. Common term for result of cutting the corpus callosum with 26 across
12. Word used to describe the process required for thought
13. Cognitive task from Wisconsin is a form of a card _____
16. Magnetoencephalogram; abbr.
19. A condition where a person ignores sensory information
22. Choking reflex that does not require thought
24. 3rd of 3 words in magnetic technique
26. See 10 across for first word
27. Like music, the right hemisphere also appreciates this stimulus
28. To examine again, as in giving a cognitive task a second time
29. An idea resulting from a set of impressions

Down

2. Mental construct that is hard to determine, though the IQ test tries
3. Term for 19 across when it occurs simultaneously with a competing stimuli
4. Word for outer brain layer, with association
5. 5 across and 25 down; abbr.
8. Most emotions are expressed in this region
9. Fancy term for hearing colors
11. Frontal association cortex is particularly necessary for this aspect of thought too
14. Factor that will eventually slow the thought process in us all
15. Type of thinking used to define terms or solve math problems
17. 2nd of 3 words in magnetic technique
18. Inability to make voluntary movements, when not paralyzed, that is.
20. Impulses used for 25 down
21. Long-term unconscious state in which higher level thought and function does not occur
23. Common lab animal used for research
25. Technique from 16 across & 24 across; abbr.

CHAPTER

16 What Happens When the Brain Misbehaves?

CHAPTER SUMMARY

After 15 chapters of describing how the brain was intended to work, this chapter focuses specifically on what happens when the brain does not follow the normal course of functioning. Many of the previous chapters touched on individual disorders and diseases that affect the brain. This chapter summarizes some *organic* and *behavioral disorders* previously discussed, and presents them as members of larger classifications of related maladies. In very general terms, neurological disorders are caused by 1) genetic errors, 2) epigenetic mechanisms, 3) progressive cell death, 4) rapid cell death, or 5) a loss of neural function and connections. The causes of behavioral-psychiatric disorders, on the other hand, are less well understood but are generally considered to have a basis in abnormal function of specific brain regions. Psychiatric disorders are also diagnosed using primarily behavioral symptoms that yield few clues to underlying structural or chemical dysfunction. Compounding the problem of determining causes for behavioral-psychiatric disorders is the overwhelming complexity of the nervous system. When considering inherent individual differences in the vast cellular organization of the brain, identifying small but potentially significant variation between affected and unaffected individuals becomes a daunting task. For example, changes as subtle as decreased dendritic spine density in one particular region of the brain can contribute to major behavioral disturbance that characterizes some disorders.

However, as challenging as the task of identifying and treating disorders appears, it is central to the goal of improving mental health. This goal is particularly noteworthy when considering that nearly one half of a sample of U.S. residents will meet the criteria for a psychiatric disorder at some point in their lives. The *Diagnostic and Statistical Manual of Mental Disorders* (*DSM*) is considered a standard tool in mental health *epidemiology* research. Now in a 5th edition (*DSM-5*), this manual provides diagnostic criteria for a wide range of psychiatric disorders. For some of these disorders technology developments, particularly in the field of brain imaging, have provided new information on affected brain regions.

The goal of studying the basis for behavioral disorders is to provide effective treatments or, ultimately, a cure. Treatments include *neurosurgical approaches* such as *deep brain stimulation* (*DBS*) that use electrical impulses to evoke adaptive behavioral responses.

Neural implants using fetal stem cells, bone marrow cells, and tumor cells are also recent neurosurgical approaches that have yielded some promising results. *Electrophysiological treatments* include *electroconvulsive therapy* (*ECT*) and *transcranial magnetic stimulation* (*TMS*). Both ECT and TMS have therapeutic potential, particularly in treating depression and schizophrenia, respectively. *Pharmacological treatments* were revolutionized nearly 50 years ago, and have developed into the most effective and preferred class of treatments for a wide range of disorders. However, pharmacological treatments are often associated with significant side effects and the potential for drug abuse. The final broad category of treatment for both neurological and psychiatric disorders is *behavioral treatment.* This category includes *behavior therapies* such as *neuropsychological therapy*, *emotional therapy*, physical activity, music therapy, *real time fMRI*, and *virtual reality therapy*.

While the causes of neurological disorders are generally well understood, treatments are often relatively primitive. Fore example, *traumatic brain injury* caused by intracranial bleeding or swelling (often resulting from a severe blow to the head) can be readily localized using advanced technology such as *magnetic resonance spectroscopy* (*MRS*). However, few effective treatments have been developed to enhance recovery in these patients. *Stroke* is a result of *ischemia* which directly damages the blood-deprived area and often indirectly affects related brain structures through a form of neural shock termed *diaschisis*. Researchers are currently developing *neuroprotectants* designed specifically to reduce cell damage associated with diaschisis. *Epilepsy* is marked by abnormally high electrical activity in the brain, resulting in seizures. *Symptomatic seizures* have a specific cause (such as a tumor, infection, etc.), while *idiopathic seizures* appear in the absence of a specific origin. *Focal seizures* have a specific point of origin from which they spread to affect a larger region of the brain (or the entire brain). *Generalized seizures* lack a focal point of origin, often occurring bilaterally. While there are a wide range of symptoms associated with seizures, three of the most common are 1) an aura prior to onset, 2) loss of consciousness, and 3) motor disturbance. Symptoms are also more prevalent and severe in *grand mal* than in *petit mal* seizures. Though specific treatments have not been developed, low doses of anesthetic agents have been moderately successful in stabilizing brain activity in some patients prone to seizures.

Multiple sclerosis (*MS*) is a disease characterized by damage to myelin and consequently disruption of neural signals in affected cells. Although it affects over a million people worldwide, the basis for this disorder is not well understood. Some researchers have proposed bacterial infection or environmental toxins as possible causes. Still other researchers believe MS is an *autoimmune disorder* and have begun developing treatment strategies targeting the immune system.

Unlike MS, which may occur at any stage of life, the vast majority of *neurodegenerative disorders* are diagnosed in the elderly. *Dementia* is a persistent syndrome of intellectual impairment. When caused by infection, exposure to toxins, vitamin deficiencies, or other specific pathological conditions, it is termed *nondegenerative dementia*. When associated with *Parkinson's disease* (PD), *Alzheimer's disease* (AD), or other diseases of aging, it is termed *degenerative dementia*. While dementia is one symptom of PD, the most notable symptoms are disturbances of motor function caused by degeneration of the substantia nigra and the subsequent loss of dopaminergic input to the striatum. Motor dysfunctions are generally categorized as *positive symptoms* or *negative symptoms*. Positive symptoms include *tremor at rest*, increased muscle tone leading to *muscular rigidity*, and patterns of abnormal *involuntary movements* termed *akasthesia*. Negative symptoms include *disorders of posture* and *disorders of equilibrium* as well as *disorders of righting* marked by difficulty standing or, in some cases, rolling over. A slowing of movements termed *akinesia* is also a common negative symptom and may contribute to symptoms such as *disturbances of speech* and *disorders of locomotion*. While the neural pathology of PD is relatively well understood, the cause of underlying neural degeneration is still unclear, with most researchers suggesting toxic effects from a yet-to-be-determined environmental factor (e.g., pollutants, insecticides, herbicides). There is no cure for PD. But treatments include the use of dopamine agonists, neurosurgery, deep brain stimulation, and fetal and stem cell implants. The effectiveness of each type of treatment varies widely, but all share the problem of significant side effects and eventual loss of effectiveness due to the progressive nature of PD.

AD is the most prevalent cause of dementia, accounting for 65 percent of all reported cases. Like PD, the cause of AD is currently unknown. Possible causes include genetics, environmental toxins, and autoimmune response, to name a few. Predominant neuroanatomical changes in AD include the development of *amyloid plaques* and *neurofibrillary tangles*, particularly in the cortex. There is no effective treatment for AD and with progression of the disease the cortex may atrophy, losing as much as one third of its original volume. One recent line of research has focused on identifying similarities between PD and AD, even suggesting a common underlying pathology. This theory of common pathology is based in part on similarities seen in behavioral features and the presence of *Lewy bodies* associated with abnormal metabolism in cells of both PD and AD patients.

As with neurological disorders, there is a wide array of psychiatric disorders. Three of the best studied and most prevalent categories of psychiatric disorders include *psychotic disorders*, *mood disorders*, and *anxiety disorders*. Schizophrenia is the most common and best understood psychotic disorder. It may be characterized by positive symptoms such as hallucinations, delusions, agitated movements, disorganized thoughts, and disorganized speech. It may also be characterized by negative symptoms such as loss of drive, lack of emotion, and lack of affect. In basic terms, patients are usually categorized as suffering from *Type I schizophrenia* (marked by predominantly positive symptoms) or *Type II schizophrenia* (marked by predominantly negative symptoms). Type I is also characterized by acute onset and generally has a better prognosis for treatment or recovery than Type II. Schizophrenia has historically been attributed to a deficiency in dopamine function. That theory, however, is currently being challenged and viewed as oversimplified. This challenge is based on recent research showing significant abnormalities in both the GABA and glutamate systems of some schizophrenics.

Mood disorders include *mania* marked by extreme and excessive euphoria and *bipolar disorder* characterized by swings from mania to *depression*. However, the best studied and most prevalent mood disorder is clinical depression. It is well established that many cases of depression can be significantly alleviated by drugs that acutely enhance serotonin and/or norepinephrine function. However, this effect does not necessarily identify the basis of the disorder. Some researchers have suggested that depression results from abnormal response to stress, particularly by the hypothalamic-pituitary-adrenal system (*HPA-axis*). More specifically, overactivation of the HPA-axis can increase circulating *cortisol* levels. Chronic elevated cortisol is correlated to depression and, in some cases, neuronal death.

Anxiety disorders represent the most commonly reported class of psychiatric disorders, affecting more than 20 percent of the U.S. population at some point in their lifetime. Imaging studies suggest these disorders may be caused by overresponsiveness of some brain areas to anxiety-provoking stimuli. The most effective treatment is generally use of drugs that enhance GABA, ultimately increasing the inhibitory tone of the brain. In some cases, selective serotonin reuptake inhibitors (generally thought of as treatment for depression) are also effective in reducing anxiety disorders. Anxiety disorders may also be reduced with *cognitive-behavior treatments*. This is particularly true in the case of obsessive-compulsive disorder and phobias.

KEY TERMS

The following is a list of important terms introduced in Chapter 16. Give the definition of each term in the space provided.

Research on Disorders

Posttraumatic stress disorder (PTSD)

Phenylketonuria (PKU)

Organic disorders

Behavioral-psychiatric disorders

Classifying and Treating Disorders

Epidemiology

Diagnostic and Statistical Manual of Mental Disorders (DSM)

Neurosurgical treatments

Deep brain stimulation (DBS)

Electroconvulsive therapy (ECT)

Transcranial magnetic stimulation (TMS)

Pharmacological treatments

Behavioral treatment

Neuropsychological therapy

Behavioral therapies

Behavior modification

Cognitive therapy

Emotional therapy

Real time fMRI

Virtual reality therapy

Traumatic Brain Injury

Magnetic resonance spectroscopy (MRS)

Stroke

Ischemia

Diaschisis

Neuroprotectants

Epilepsy

Symptomatic seizures

Focal seizures

Generalized seizures

Aura

Grand mal

Petit mal

Multiple sclerosis (MS)

Autoimmune disorder

Neurodegenerative disorders

Dementia

Nondegenerative dementia

Parkinson's disease

Alzheimer's disease

Degenerative dementia

Positive symptoms

Negative symptoms

Tremor at rest

Muscular rigidity

Involuntary movements

Akasthesia

Disorders of posture

Disorders of equilibrium

Disorders of righting

Akinesia

Disturbances of speech

Disorders of locomotion

Amyloid plaques

Neurofibrillary tangles

Lewy bodies

Understanding Psychiatric Disorders

Psychotic disorders

Mood disorders

Anxiety disorders

Positive symptoms

Negative symptoms

Type I schizophrenia

Type II schizophrenia

Mania

Bipolar disorder

Depression

HPA-axis

Anxiety disorders

Cognitive-behavior treatments

PRACTICE TEST

Multiple-Choice Questions

Answer each of the following multiple-choice questions with the best possible answer based on information from your text.

1. Freud proposed that the "id" was the root of emotional behaviors. Currently, we believe that emotional behaviors are largely mediated by the ________________, which could be thought of in psychoanalytic terms as a modern day "id".
   A. spinal cord
   B. frontal lobe
   C. limbic system
   D. cerebellum
   E. pituitary gland

2. What is the percentage of Americans who meet the criteria for a psychiatric disorder at some point in their lives?
   A. less than 10 percent
   B. 10–20 percent
   C. 20–30 percent
   D. 30–40 percent
   E. greater than 40 percent

3. Which of the following research techniques has provided significant evidence of presymptomatic abnormal brain function in children who eventually develop schizophrenia?
   A. neurochemical techniques
   B. postmortem brain tissue analysis
   C. MRI scans
   D. psychotherapy
   E. All of the answers are correct.

4. Which of the following was *not* listed as a basic category of treatments for brain and behavior disorders?
   A. neurosurgical
   B. homeopathic
   C. pharmacological
   D. electrophysiological
   E. behavioral

5. Which of the following would *not* be considered a treatment method for brain dysfunction?
   A. ECT
   B. DBS
   C. TMS
   D. Stem cell implants
   E. fMRI

6. Though pharmacological therapies of psychiatric disorders are considered among the most effective treatments, what tops the list of problems with such drug treatments?
   A. cost
   B. side effects
   C. potential for addiction
   D. FDA restrictions
   E. public skepticism about drug use

7. Drug treatment for which of the following disorders is most likely to induce tardive dyskinesia?
   A. Parkinson's disease
   B. depression
   C. anxiety disorder
   D. Alzheimer's disease
   E. schizophrenia

8. Which of the following behavioral treatments would be most likely to employ systematic desensitization as a treatment technique?
   A. behavior therapy
   B. cognitive therapy
   C. psychotherapy
   D. All of the answers are correct.
   E. None of the answers is correct.

9. Which of the following most accurately describes the basis for a stroke?
   A. blockage of a blood vessel
   B. bleeding from a blood vessel
   C. an interruption of blood flow to a region of the brain
   D. an interruption of electrical activity in the brain
   E. trauma to the brain, usually caused by a blow to the head

10. *Diaschisis* is a term used to describe which of the following?
    A. abnormal movements associated with Parkinson's disease
    B. cognitive deficits associated with Alzheimer's disease
    C. neural shock associated with stroke
    D. high-level electrical activity associated with seizures
    E. loss of myelin associated with multiple sclerosis

11. Seizures caused by abnormal activity associated with a brain tumor would most appropriately be described as which of the following?
    A. idiopathic
    B. symptomatic
    C. grand mal
    D. petit mal
    E. tumoric

12. Multiple sclerosis (MS), could be described as a neurological disorder caused by which of the following?
    A. genetic error
    B. loss of myelin
    C. progressive cell death
    D. rapid cell death
    E. None of the answers is correct.

13. Which of the following is the most common course of treatment for epilepsy?
    A. daily treatment with ECT
    B. high doses of antipsychotic drugs
    C. low doses of antidepressants
    D. low doses of stimulants
    E. low doses of anesthetics

14. Which of the following is *not* a characteristic of multiple sclerosis?
    A. It is most prevalent in Northern Europe.
    B. It is frequently inherited.
    C. It affects more women than men.
    D. It is one of the most common structural diseases of the nervous system.
    E. It is rare in tropical climates.

15. Which of the following is true of akathesia?
    A. It is a symptom of Alzheimer's disease.
    B. It is categorized as a positive symptom.
    C. It is a Greek word meaning "little bad."
    D. It is particularly associated with cognitive function.
    E. All of these statements are true of akathesia.

16. Which of the following is the most prevalent cause of dementia?
    A. Korsakoff's syndrome
    B. Parkinson's disease
    C. Huntington's chorea
    D. Alzheimer's disease
    E. Wilson's disease

17. Which of the following is typically found in the brain of a patient with Alzheimer's disease?
    A. neurofibrillary tangles
    B. Lewy bodies
    C. amyloid plaques
    D. cortical atrophy
    E. All of these are found in the brain of a patient with Alzheimer's disease.

18. Which of the following is *not* a feature of Type II schizophrenia?
    A. poor prognosis
    B. poor response to neurolepics
    C. acute onset
    D. enlarged ventricles
    E. cortical atrophy

19. Which of the following stress hormones, associated with development of mood disorders, is increased when HPA-axis activity is increased?
    A. cortisol
    B. adrenaline
    C. insulin
    D. testosterone
    E. antidiuretic hormone

20. Anxiety disorders are typically treated with benzodiazepines. However, some level of success has also been achieved treating anxiety disorders with which of the following?
    A. selective serotonin reuptake inhibitors (SSRIs)
    B. antipsychotics
    C. psychomotor stimulants
    D. nicotine
    E. alcohol

Short-Answer Questions

Answer each of the following questions with a brief but complete written answer based on information from your text.

1. The causes of many neurological disorders are, at least in a general sense, well enough understood to create categories for these disorders. Give two examples of disorders from each of the following three categories: genetic error, progressive cell death, rapid cell death.

2. Phenylketonuria (PKU) represents a true success story in terms of neurological disorder research. Briefly describe the cause of PKU, its effects if left untreated, and the treatment course for an infant diagnosed with PKU.

3. Seizures have been used for nearly 100 years to treat some types of psychiatric disorders. Originally this form of therapy utilized insulin injections to evoke seizures. ECT eventually replaced insulin therapy, but is now being phased out in favor of a newer method. Name the latest method being used for therapeutic purposes that is replacing the use of ECT. Briefly explain the principles of this technique and why it has gained favor among physicians and patients.

4. Briefly describe the three developments that led to a literal revolution in pharmacological treatment of behavioral disorders.

5. List the three symptoms (noted in the text) that are considered common to a variety of epileptic episodes. Give a brief explanation or example of each symptom.

6. There is currently no cure for Parkinson's disease; however, pharmacological treatment may be effective in reducing symptoms. Briefly describe the two primary objectives of pharmacological treatment for Parkinson's disease.

7. Briefly describe the features of Type I and Type II schizophrenia.

8. Briefly explain how chronic stress might contribute to development of clinical depression. Include in your answer the HPA-axis, cortisol, and the concept of critical periods during development.

9. Briefly describe how cognitive-behavior therapy might be used to treat a specific phobia.

10. Give example of when "misbehavior" of the brain might not be considered a strictly dysfunctional condition.

Matching Questions

Complete each of the following matching questions based on information from your text.

1. Even the best treatments for neurological dysfunction have drawbacks. Match the following treatments with the most prevalent associated drawback.

A. Deep brain stimulation	_____ To date, has produced only modest improve ments in patients
B. Stem cell implants	_____ Patients often relapse when treatment is "turned off"
C. ECT	_____ Tends to produce significant memory loss
D. Pharmacological treatments	_____ Potential for significant long- and short-term side effects

2. Match the following abbreviation with the appropriate description.

MRS	_____ American Psychiatric Association publication
PKU	_____ Genetic disorder treated with dietary restrictions
DBS	_____ Neurosurgical approach to treating Parkinson's disease
DSM	_____ Most recently developed seizure therapy method
TMS	_____ Can identify changes in specific markers of neural function

3. Parkinson's disease is marked by both positive and negative symptoms. Identify each of the following symptoms as positive (+) or negative (−).

_____ Tremor at rest
_____ Akinesia
_____ Oculogyric crisis
_____ Muscular rigidity
_____ Disorders of posture
_____ Disorders of locomotion

4. Match the following symptoms to the appropriate neurological disorder.

A. Akathesia	_____ Epilepsy
B. Ischemia	_____ Parkinson's disease
C. Grand mal attack	_____ Alzheimer's disease
D. Amyloid plaques	_____ Stroke

5. Correctly match each of the following disorders to the appropriate classification.

A. Psychotic disorders	_____ Posttraumatic stress disorder
B. Mood disorders	_____ Schizophrenia
C. Anxiety disorders	_____ Bipolar disorder
	_____ Obsessive-compulsive disorder
	_____ Mania

Diagrams

1. Incidence rates for head trauma differ between males and females. The accompanying diagram shows reports of head traumas between 1965 and 1974 in Olmsted County, MN. Identify the sex represented by this diagram and draw a line approximating data for the other sex.

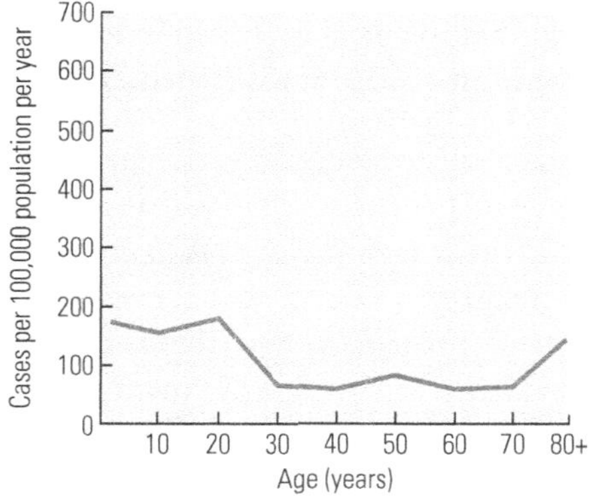

2. Brain dysfunction following stroke results not from a single event, but from a sequential cascade of events. Correctly label the events on the accompanying time-sequence graph.

Recovery
Inflammation
Ionic changes
Proteins
Second messengers
mRNA

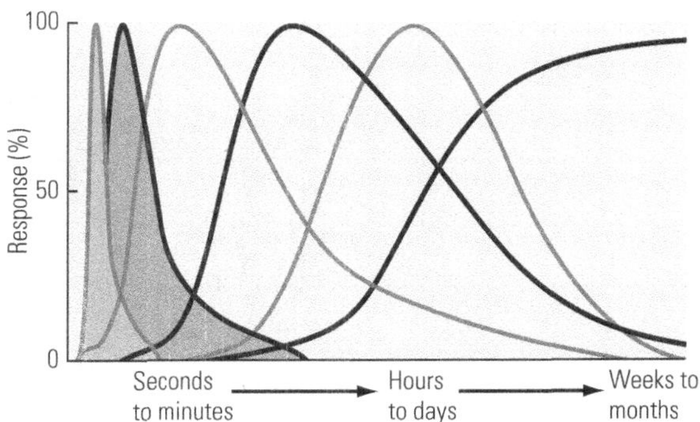

3. During a grand mal seizure, there is a predictable sequence of three distinct EEG patterns that indicate: 1) Onset of attack, 2) Clonic phase, 3) Coma period after seizure. Complete the following diagram by sketching an approximation of the EEG recordings that would be seen in these three stages.

Normal EEG Activity	Onset of Attack	Clonic Phase	Coma

4. The accompanying diagram indicates what a cell might look like in the early stages of Alzheimer's disease. Draw a representative diagram to indicate how this cell differs anatomically from an unaffected cell. Draw a second diagram to indicate how this cell differs anatomically from a cell in the advanced stages of Alzheimer's disease.

Early Alzheimer's disease

5. The accompanying diagram depicts a hippocampal neuron arrangement in an unaffected individual. Draw a representative diagram to indicate how the arrangement of these cells differs from that of cells in the hippocampus of an individual with schizophrenia.

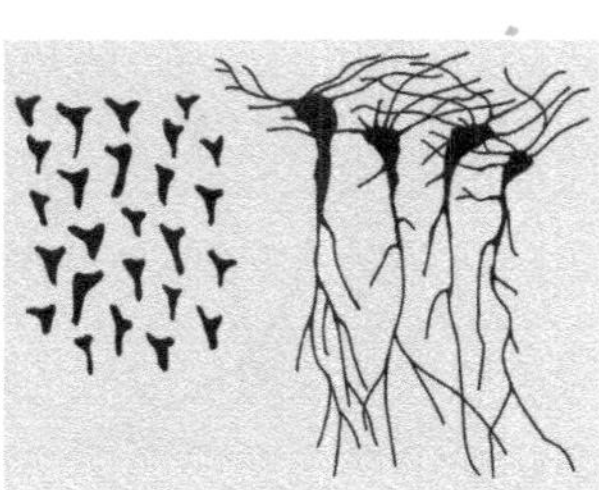

The Web

Consider using the following Web sites for additional information on some of the topics from this chapter:

1. Epilepsy foundation: www.epilepsyfoundation.org/
2. Brain and Behavior Research Foundation: www.narsad.org/
3. American Stroke Association: www.strokeassociation.org
4. Parkinson's disease foundation: www.pdf.org/
5. Alzheimer's Disease Education & Referral Center: www.nia.nih.gov/alzheimers

CROSSWORD PUZZLE

Across

1. A systematic form of behavior therapy
3. Term for nonspecific form; with 16 down
10. Classification word for schizophrenia
11. _____ symptoms of schizophrenia include hallucinations and delusions
14. Myelin degenerating disease; abbr.
15. The "M" in 20 across
17. Term for sensation that precedes a seizure
18. Positive Parkinson's symptom or a small earthquake
19. Cellular indicator of mitochondrial dysfunction; with 13 down
20. American Psychiatric Association publication used to classify behavior disorders; abbr.
21. Oversensitive HPA _____ may contribute to depression
22. Type of brain stimulation sometimes used to treat Parkinson's

Down

1. Dementia type associated with Parkinson's or Alzheimer's; e.g.
2. The poor prognosis classification; with 10 across
4. Disease responsible for largest percentage of dementia reports
5. Type of seizure or a dress size
6. Plural form of 13 down
7. Therapy using electricity; abbr.
8. Disorder combining mania and depression
9. Result of 7 down and 18 down therapies
12. Predecessor was the MRI
13. "Soma" in Latin
14. May be "grand" or 5 down
15. Disorder classification for depression, mania, and 8 down
16. Specific forms include phobias and panic attacks
18. Therapy using magnetic fields; abbr.
23. Genetic disorder treated with dietary restrictions; abbr.

Answers

Chapter 1: What Are the Origins of Brain and Behavior?

Multiple-Choice Questions

1. E (p. 3)
2. B (p. 4)
3. E (p. 4)
4. D (p. 5)
5. C (p. 6)
6. B (p. 7)
7. C (p. 7)
8. C (p. 7)
9. C (p. 8)
10. E (p. 11)
11. C (p. 13)
12. C (p. 15)
13. E (p. 15)
14. A (p. 16)
15. C (p. 21)
16. D (p. 22)
17. C (p. 23)
18. E (p. 26)
19. A (p. 29)
20. C (p. 29)

Short-Answer Questions

1. (p. 3)
2. (p. 4)
3. (p. 6)
4. (p. 7)
5. (p. 8)
6. (p. 15)
7. (p. 23)
8. (p. 25)
9. (p. 26)
10. (p. 29)

Matching Questions

1. _C_ Mammal
   _E_ Great Apes
   _D_ Primates
   _A_ Animal
   _F_ Human
   _B_ Chorates

2. _D_ Materialism
   _C_ Dualism
   _A_ Genes
   _B_ Mentalism

3. _D_ Most recently evolved species
   _C_ Co-existed with early modern-day humans
   _A_ First to walk upright
   _B_ Known for their use of tools

4. _1_ Rat
   _5_ *Homo sapiens*
   _2_ Elephant
   _3_ Chimpanzee
   _4_ Dolphin

Crossword Puzzle

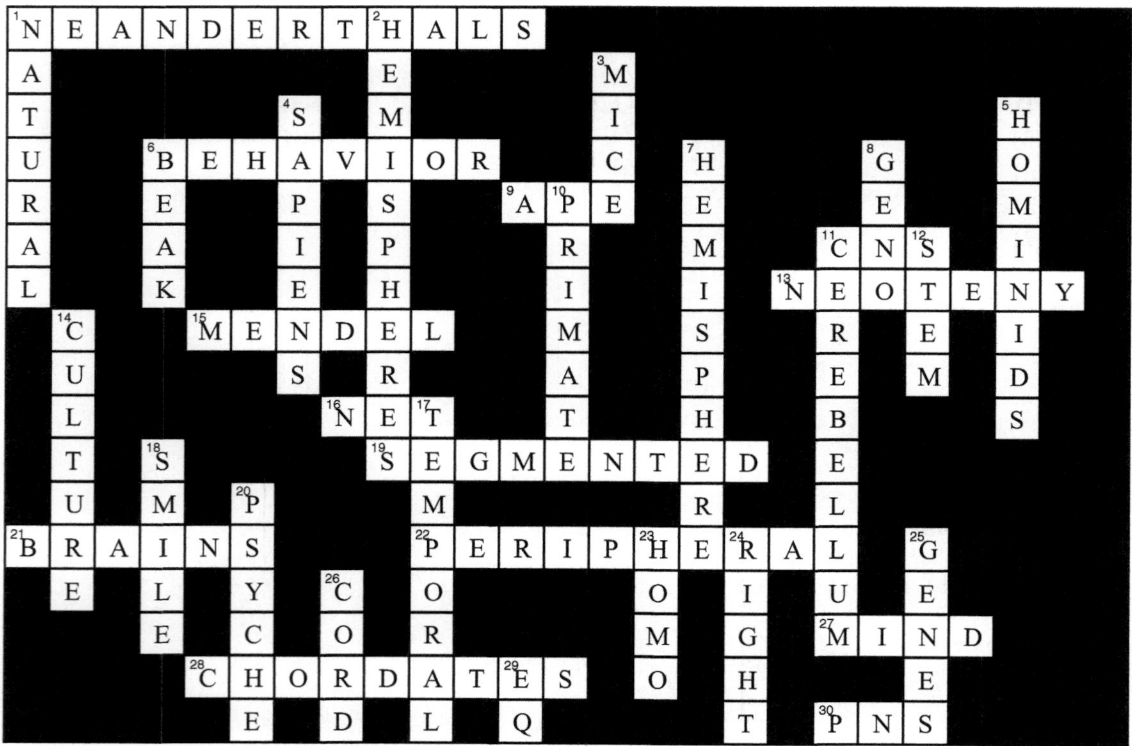

Chapter 2: How Does the Nervous System Function?

Multiple-Choice Questions

1. C (p. 37)
2. B (p. 37)
3. A (p. 38)
4. C (p. 40)
5. C (p. 41)
6. C (p. 47)
7. D (p. 48)
8. E (p. 52)
9. E (p. 53)
10. B (p. 53)
11. C (p. 53)
12. B (p. 54)
13. E (p. 55)
14. E (p. 58)
15. C (p. 58)
16. B (p. 60)
17. A (p. 61)
18. B (p. 64)
19. B (p. 66)
20. E (p. 68)

Short-Answer Questions

1. (p. 38)
2. (p. 40)
3. (p. 52)
4. (p. 54)
5. (p. 56)
6. (p. 59)
7. (p. 60)
8. (p. 62)
9. (p. 67)
10. (p. 70)

Matching Questions

1. A Anterior
   B Posterior
   F Ventral
   C Medial
   E Superior
   F Inferior
   B Caudal
   A Rostral

2. B Peripheral nervous system (PNS)
   C Sympathetic nervous system
   A Central nervous system (CNS)
   D Internal nervous system

3. D Hypothalamus
   B Ventricles
   E Neocortex
   C Cerebellum
   A Corpus callosum

4. 1 Cortex
   3 Midbrain
   2 Diencephalon
   4 Hindbrain

5. C Basal ganglia
   E May affect many regions of the brain
   D Pia mater and arachnoid layer
   A Limbic system
   B Facial nerves

Diagrams

1.

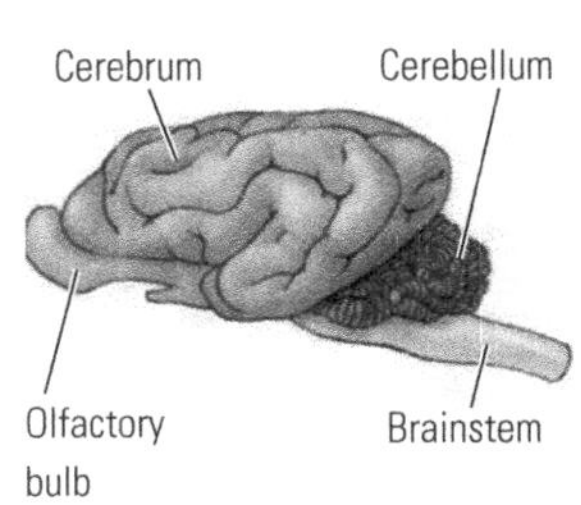

2.

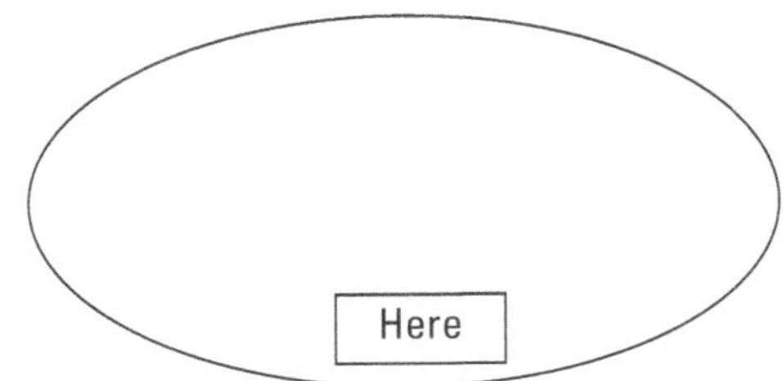
Here

3.

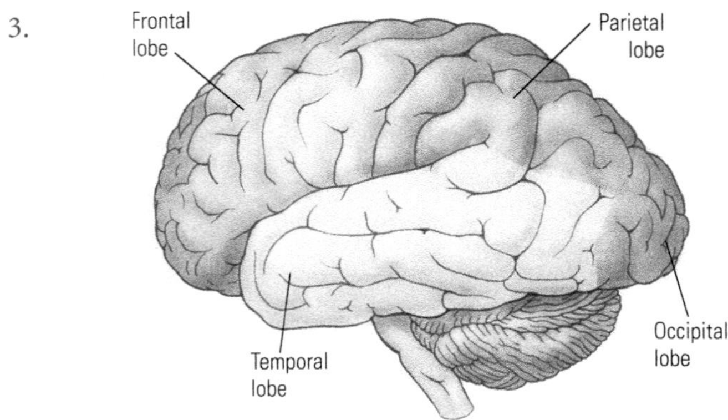
Frontal lobe
Parietal lobe
Occipital lobe
Temporal lobe

4.

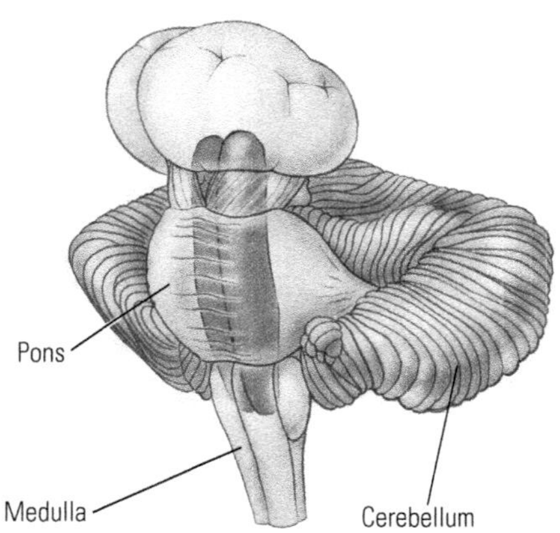
Pons
Medulla
Cerebellum

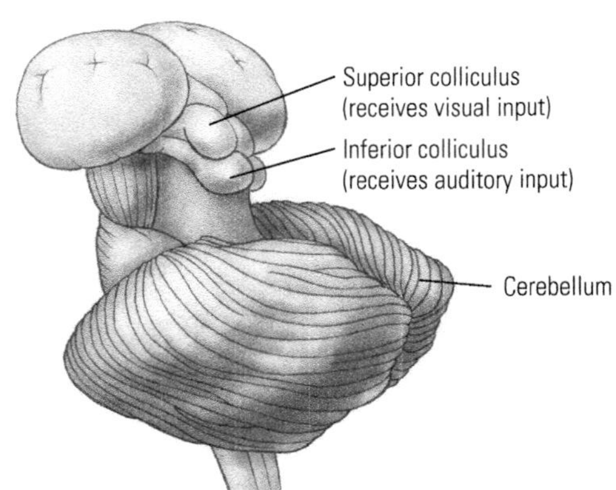
Superior colliculus (receives visual input)
Inferior colliculus (receives auditory input)
Cerebellum

5.

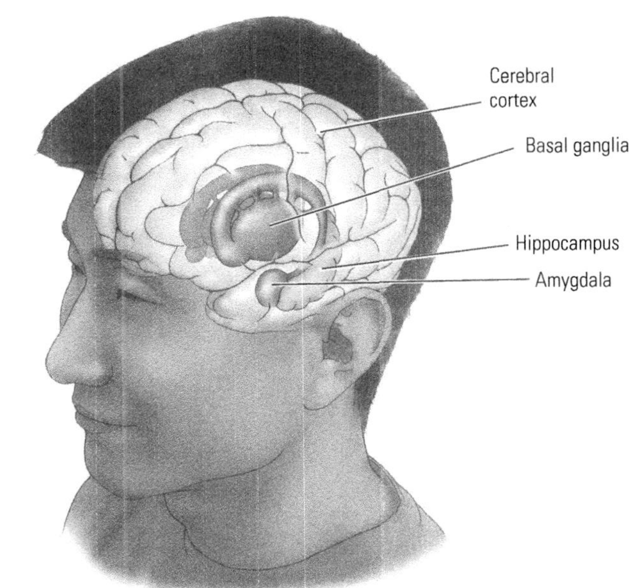
Cerebral cortex
Basal ganglia
Hippocampus
Amygdala

6.

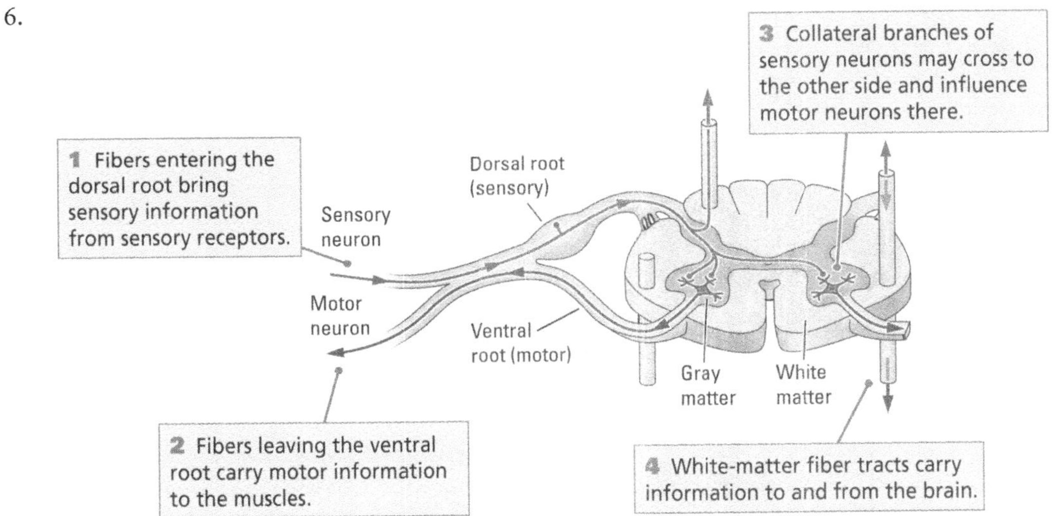
1 Fibers entering the dorsal root bring sensory information from sensory receptors.
Sensory neuron
Motor neuron
Dorsal root (sensory)
Ventral root (motor)
3 Collateral branches of sensory neurons may cross to the other side and influence motor neurons there.
Gray matter
White matter
2 Fibers leaving the ventral root carry motor information to the muscles.
4 White-matter fiber tracts carry information to and from the brain.

7.

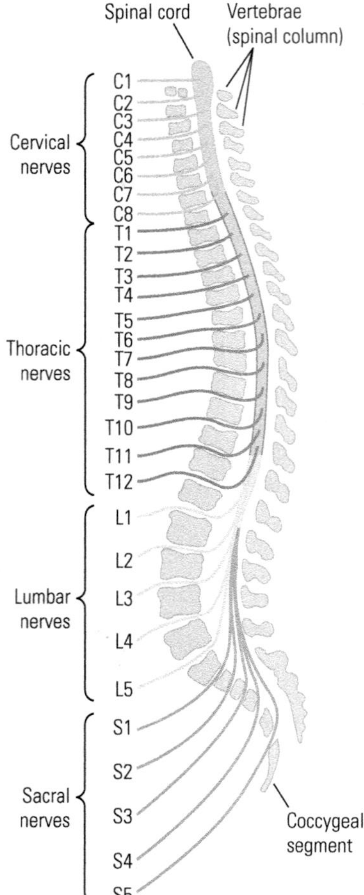
Spinal cord
Vertebrae (spinal column)
Cervical nerves
C1
C2
C3
C4
C5
C6
C7
C8
Thoracic nerves
T1
T2
T3
T4
T5
T6
T7
T8
T9
T10
T11
T12
Lumbar nerves
L1
L2
L3
L4
L5
Sacral nerves
S1
S2
S3
S4
S5
Coccygeal segment

8.

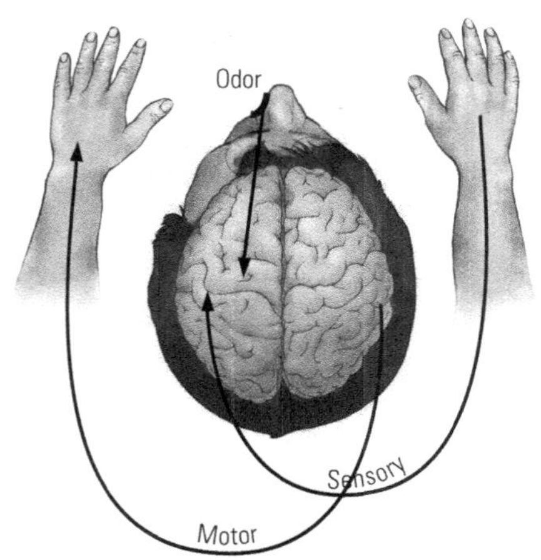
Odor
Sensory
Motor

Crossword Puzzle

[1] S	A	[2] G	I	T	T	[3] A	L		[4] C		[5] R					[6] V							
U		Y				X			[7] N	E	O	C	O	R	T	E	X		[8] F		[9] S		
B		R		[10] G	L	O	B	U	S		O					N		[11] C	O	R	T	E	X
C		I				N					T			[12] C		T			R		R		
O					[13] C	S	F							O		R			E		O		
R					O					[14] D		[15] S	U	L	C	I			B		K		
T			[16] H	I	N	D	[17] B	R	A	I	N			L		C			R		E		
I					T		A			E				I		L			A				
[18] C	E	R	E	B	R	O	S	P	I	N	A	L		C		E			I				
A					A		A			C			[19] S	U	B	S	T	A	N	T	I	[20] A	
L					L		[21] L	O	B	E		[22] L		L								M	
	[23] D	O	R	S	A	L				P		I		U								Y	
					T					[24] H	E	M	I	S	P	H	E	R	E	S		G	
					E					A		B										D	
		[25] P		[26] F	R	O	N	T	A	L		I										A	
		I			A					O		[27] C	O	R	P	U	S					L	
	[28] G	A	N	G	L	I	A		[29] P	N	S											A	

Chapter 3: What Are the Functional Units of the Nervous System?

Multiple-Choice Questions

1. E (p. 77)
2. C (p. 77)
3. D (p. 78)
4. A (p. 79)
5. D (p. 82)
6. B (p. 84)
7. D (p. 85)
8. C (p. 85)
9. E (p. 89)
10. A (p. 91)
11. D (p. 92)
12. E (p. 92)
13. C (p. 94)
14. B (p. 94)
15. E (p. 95)
16. A (p. 97)
17. E (p. 98)
18. D (p. 98)
19. D (p. 100)
20. D (p. 103)

Short-Answer Questions

1. (p. 79)
2. (p. 82)
3. (p. 83)
4. (p. 84)
5. (p. 86)
6. (p. 89)
7. (p. 93)
8. (p. 96)
9. (p. 100)
10. (p. 102)

Matching Questions

1. _C_ Soma
   _A_ Terminal
   _D_ Axon
   _B_ Dendrite

2. _F_ Single short axon and single short dendrite
   _E_ Dendrite connected directly to axon
   _D_ Many dendrites extend directly from soma
   _B_ Long axon with two sets of dendrites extend from soma
   _C_ Many dendritic branches form a fan shape
   _A_ Extensive dendrites, large soma, long axon to muscle

3.

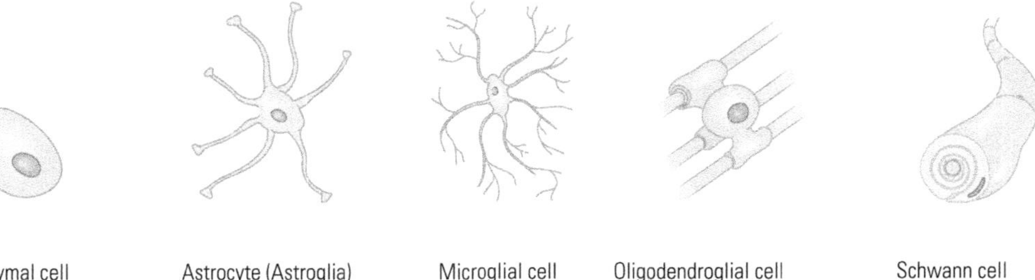

4. _A_ Cell membrane
_C_ Endoplasmic reticulum
_B_ Nucleus
_E_ Microtubules
_D_ Golgi bodies
_G_ Lysosomes
_F_ Mitochondria

5. _B_ Caused by recessive allele
_A_ Caused by dominant allele
_D_ No known genetic contribution
_C_ Caused by additional chromosome

Diagrams

1.

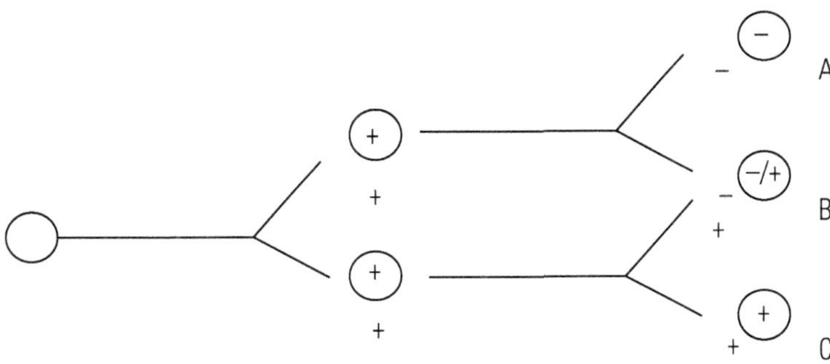

2.

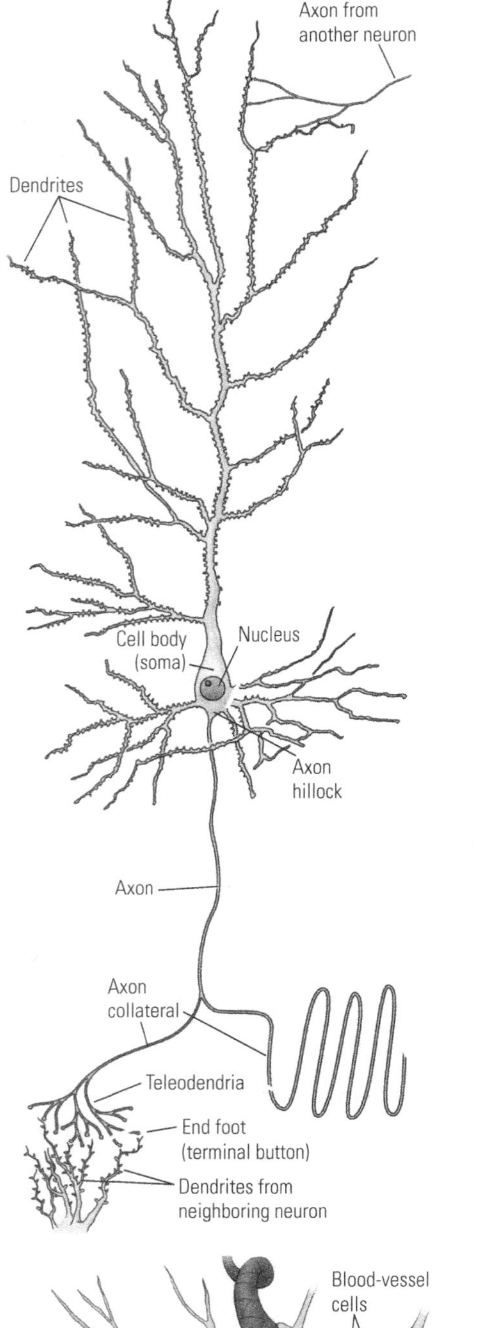
Axon from
another neuron
Dendrites
Cell body
(soma)
Nucleus
Axon
hillock
Axon
Axon
collateral
Teleodendria
End foot
(terminal button)
Dendrites from
neighboring neuron

3.

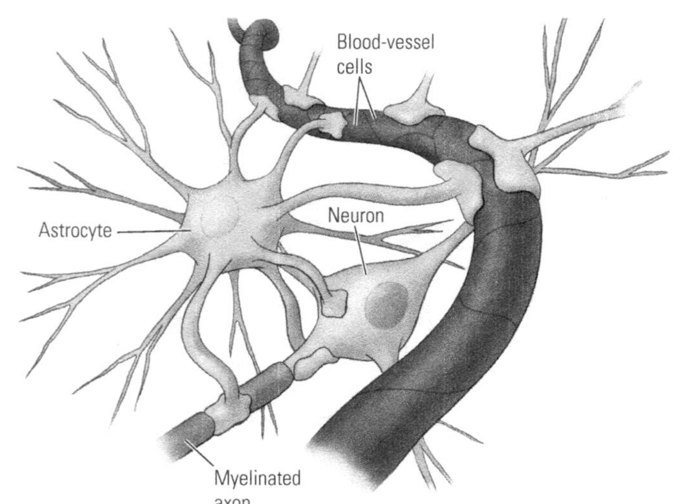
Blood-vessel
cells
Astrocyte
Neuron
Myelinated
axon

4.

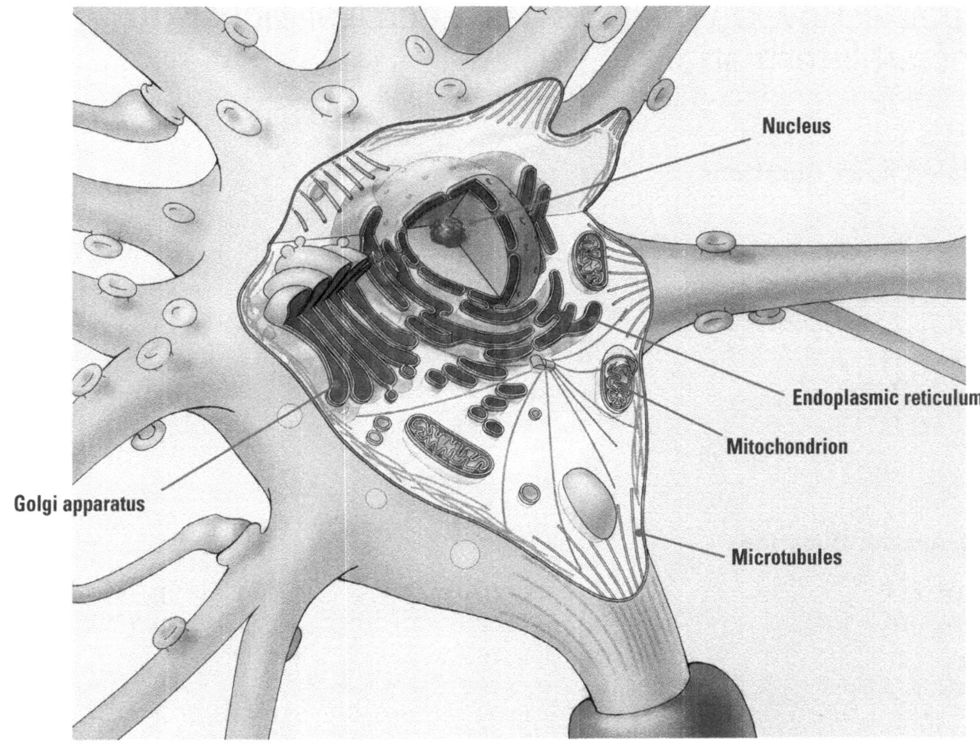

5.

HN + NN	HN + HN	HH + NN	TN + NN	TN + TN	TT + NN
.5	.75	1.0	0.0	.25	0.0

Crossword Puzzle

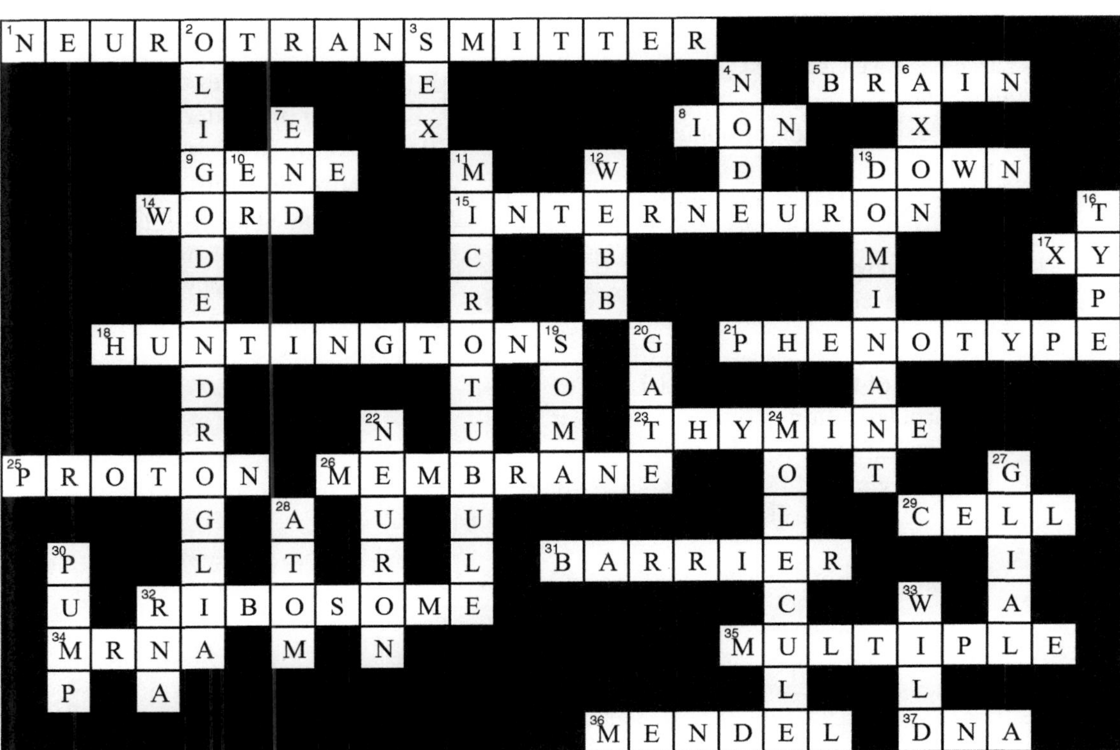

Chapter 4: How Do Neurons Use Electrical Signals to Transmit Information?

Multiple-Choice Questions

1. C (p. 113)
2. C (p. 112)
3. B (p. 113)
4. A (p. 114)
5. E (p. 115)
6. A (p. 117)
7. E (p. 117)
8. D (p. 119)
9. E (p. 120)
10. A (p. 120)
11. E (p. 121)
12. C (p. 123)
13. B (p. 124)
14. C (p. 127)
15. A (p. 127)
16. D (p. 129)
17. E (p. 131)
18. D (p. 132)
20. D (p. 135)
19. C (p. 136)

Short-Answer Questions

1. (p. 113)
2. (p. 115)
3. (p. 117)
4. (p. 118)
5. (p. 119)
6. (p. 122)
7. (p. 124)
8. (p. 126)
9. (p. 127)
10. (p. 129)

Matching Questions

1. _+_ Postassium
   _+_ Sodium
   _+_ Calcium
   _–_ Chloride
   _–_ Intracellular proteins
2. _I_ Large negatively charged protein molecules
   _I_ Higher concentration of potassium ions
   _E_ Higher concentration of sodium ions
   _I_ Negative charge
   _E_ Higher concentration of chloride ions
3. _IS_ Initation of the action potential
   _EP_ Hyperpolarization
   _IS_ Depolarization
   _EP_ Refractory period
4. _B_ Insensitivity to the chemical messages
   _B_ Autoimmune disorder
   _C_ Produces abnormal EEG
   _A_ Causes degeneration of motor neurons
   _C_ Neurons fire synchronously
5. _AP_ Generally changes membrane potential by greater than 50mV
   _GP_ May move in a positive or negative direction
   _AP_ Is initiated at a threshold charge
   _AP + GP_ May occur when sodium channels open
   _AP_ Concludes with a refractory period

Diagrams

1.

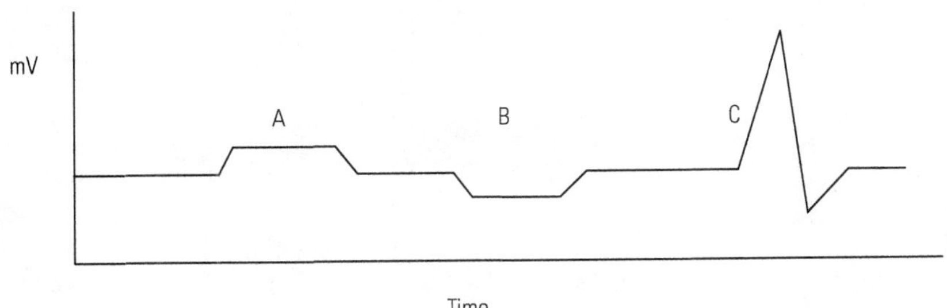

2.

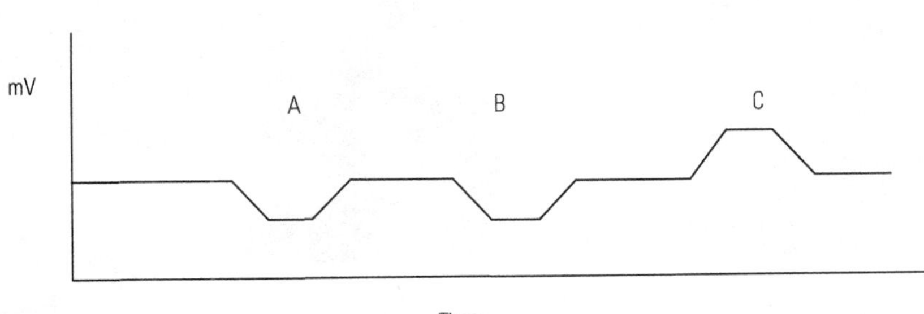

3.

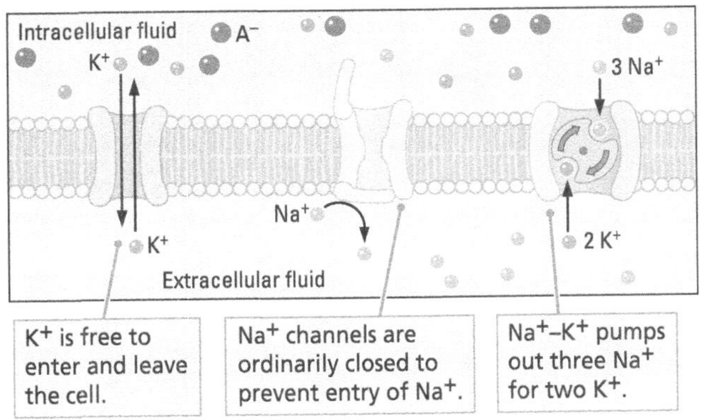

4.

5. Neuron B

Crossword Puzzle

1 O	2 S	C	I	L	L	O	S	C	3 O	P	E			4 P	O	T	A	5 S	S	I	U	M		
	P								L					O				P						
	I		6 E	L	7 E	C	T	R	I	C	A	L		T			8 R	A	N	V	I	9 E	R	
	N				P				G					E				T				E		
	A				S			10 P	O	S	T	S	Y	N	A	P	T	I	C			G		
	L				P		11 E	N	D					T				A						
		12 O	U	13 T				14 S	E	N	15 S	I	T	I	V	E		16 L	17 A	W				
				H					N		T			A					X		18 R			
	19 P			R					D		I		20 E	L	E	C	21 T	R	O	D	E			
22 H	U	X	L	E	Y				R		M			S			W		N		S			
	M			S		23 A	B	S	O	L	U	T	E				O				T			
	P			H					G		L										24 I	P	S	P
				O				25 G	L	I	A		26 G		27 S	C	H	W	A	N	N			
		28 V	O	L	29 T				I		T		I								G			
		I		30 D	E	P	O	L	A	R	I	Z	A	T	I	O	N							
		A			A						N		N											
									31 N	E	G	A	T	I	V	E								

Chapter 5: How Do Neurons Use Electrochemical Signals to Communicate and Adapt?

Multiple-Choice Questions

1. B (p. 140)
2. B (p. 141)
3. D (p. 144)
4. E (p. 145)
5. D (p. 147)
6. A (p. 148)
7. B (p. 148)
8. D (p. 148)
9. C (p. 149)
10. A (p. 152)
11. C (p. 152)
12. B (p. 153)
13. D (p. 154)
14. E (p. 156)
15. E (p. 158)
16. C (p. 161)
17. D (p. 164)
18. D (p. 166)
19. C (p. 168)
20. C (p. 168)

Short-Answer Questions

1. (p. 144)
2. (p. 146)
3. (p. 147)
4. (p. 150)
5. (p. 155)
6. (p. 155)
7. (p. 158)
8. (p. 158)
9. (p. 164)
10. (p. 165)

Matching Questions

1. B Main excitatory transmitter
   A Found at neuromuscular junction
   C Main inhibitory transmitter
   E Synthesized as a gas
   D Used for pain reduction

2. D, B Depression
   A, B Schizophrenia
   A Drug abuse
   A Parkinson's disease
   C Alzheimer's disease
   D Manic behavior

3. C Aplysia sprayed with water
   A MPTP esposed humans
   D Drosophila with rutabaga mutation
   B Heart muscle fibers

4. I Binding site attached directly to membrane pore
   M Changes cell activity through a series of steps
   M Associated with G-proteins
   I Mediates rapid changes in membrane voltage
   I Structurally similar to voltage-sensitive channel
   M Often utilizes second messenger systems

5. P Rest-and-digest
   S Heart rate increase
   P Cholinergic neurons
   S Adrenergic neurons
   S Fight-or-flight

Diagrams

1.

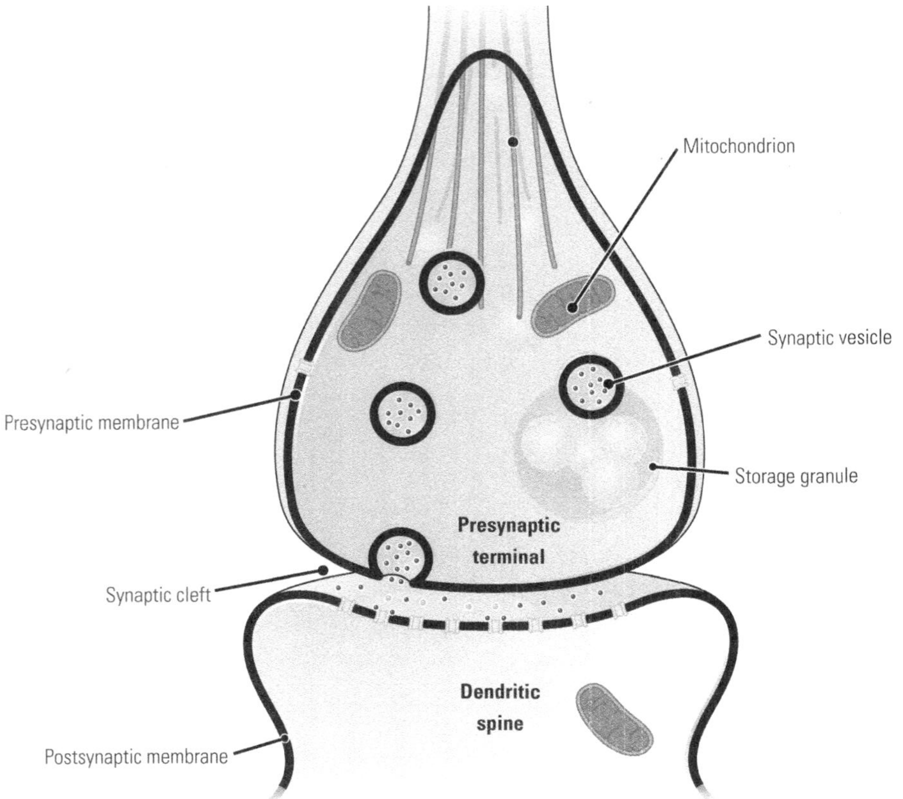
Mitochondrion
Synaptic vesicle
Presynaptic membrane
Storage granule
Presynaptic terminal
Synaptic cleft
Dendritic spine
Postsynaptic membrane

2.

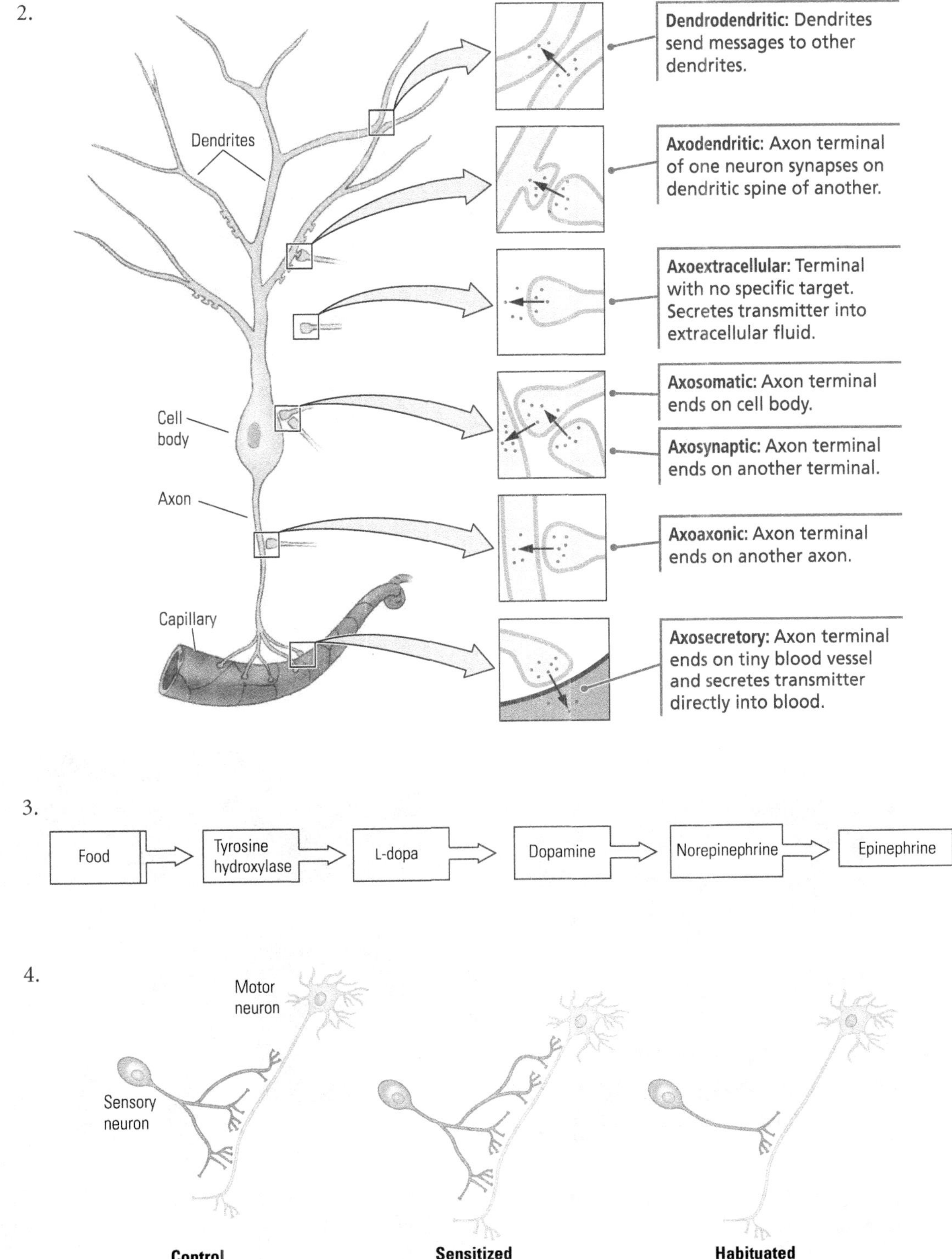
Dendrites
Cell body
Axon
Capillary
Dendrodendritic: Dendrites send messages to other dendrites.
Axodendritic: Axon terminal of one neuron synapses on dendritic spine of another.
Axoextracellular: Terminal with no specific target. Secretes transmitter into extracellular fluid.
Axosomatic: Axon terminal ends on cell body.
Axosynaptic: Axon terminal ends on another terminal.
Axoaxonic: Axon terminal ends on another axon.
Axosecretory: Axon terminal ends on tiny blood vessel and secretes transmitter directly into blood.
3.
Food
Tyrosine hydroxylase
L-dopa
Dopamine
Norepinephrine
Epinephrine
4.
Motor neuron
Sensory neuron
Control
Sensitized
Habituated

5.

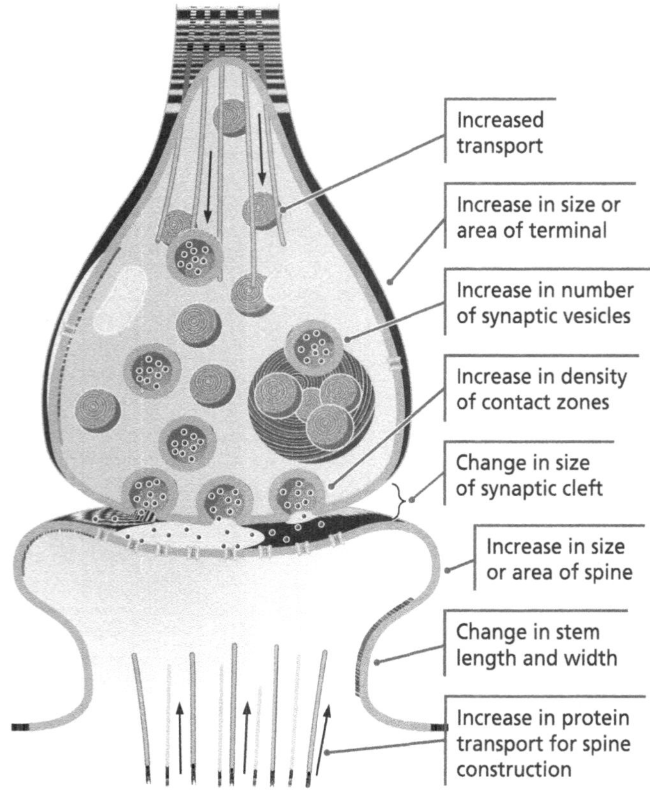

Crossword Puzzle

[1] N	O	R	E	P	I	N	E	P	H	R	I	N	E											
E																					[2] G			
U				[3] C					[4] H									[5] L			A			
R				L					[6] A	[7] C	E	T	A	T	E			E			M			[8] S
[9] O	X	I	D	E			[10] G	A	B	A						[11] G	L	U	[12] T	A	M	A	T	E
T				F					I										R		A			C
R				T			[13] M	[14] E	T			[15] E					[16] R		A				[17] D	O
A								[18] P	U	B		P			[19] D		E		N				Y	N
N				[20] L					A		[21] N	I	G	R	O		T		S				S	D
[22] S	E	N	S	I	[23] T	I	Z	A	T	I	O	N			P		R		P				K	
M				[24] E	R				I			E			A		O		O				I	
I					E		[25] T	W	O			P			M		G		R				N	
T			[26] B		M				[27] N	[28] A	[29] C	H			I		R		T				E	
T			E		O					[30] C	A	R	B	O	N		A		E				S	
E		[31] S	T	O	R	A	G	E		H		I			[32] E	N	D	O	R	P	H	I	N	S
R			A		S							N					E						A	
S												E												

Chapter 6: How Do Drugs and Hormones Influence the Brain and Behavior?

Multiple-Choice Questions

1. E (p. 173)
2. C (p. 173)
3. B (p. 174)
4. C (p. 176)
5. D (p. 177)
6. B (p. 178)
7. E (p.179)
8. C (p. 182)
9. D (p. 183)
10. E (p. 185)
11. A (p. 186)
12. E (p. 186)
13. B (p. 188)
14. A (p. 190)
15. C (p. 191)
16. D (p. 193)
17. B (p. 195)
18. B (p. 197)
19. B (p. 202)
20. E (p. 206)

Short-Answer Questions

1. (p. 174)
2. (p. 178)
3. (p. 185)
4. (p. 186)
5. (p. 190)
6. (p. 193)
7. (p. 195)
8. (p. 196)
9. (p. 198)
10. (p. 206)

Matching Questions

1. _5_ Oral (eaten)
   _4_ Topical (applied to the surface of the skin or mucus)
   _2_ Intravenous (injected into a vein)
   _1_ Intracranial (injected into the brain)
   _3_ Inhalation (smoked)

2. agonists — block presynaptic reuptake of neurotransmitter from synapse
   agonists — block enzyme that breaks down neurotransmitter
   antagonists — block postsynaptic receptor site
   antagonists — block release of neurotransmitter from presynaptic terminal
   agonists — increase effectiveness of neurotransmission
   antagonists — decrease effectiveness of neurotransmission

3. _B_ Block dopamine receptor sites
   _A_ Affects $GABA_A$ receptor
   _E_ Increases cyclic AMP
   _D_ Among the most potent analgesics
   _C_ Contributed to development of Prozac

4. _E_ Fermented and distilled
   _C_ Coca plant
   _B_ Hemp plant
   _D_ Poppy seeds
   _A_ Peyote cactus sugars

5. _C_ More common in women than in men
   _A_ Poor nutrition may increase symptoms
   _A_ Associated with underdeveloped brain
   _C_ Sometimes treated with electroconvulsive therapy
   _B_ May be co-diagnosed with schizophrenia

Diagrams

1.

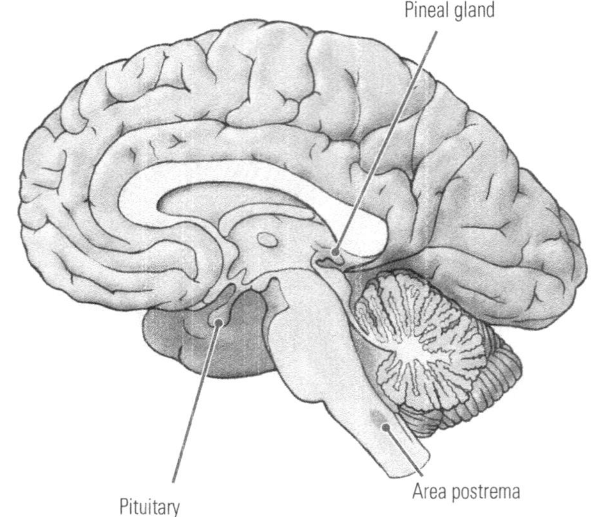

2.

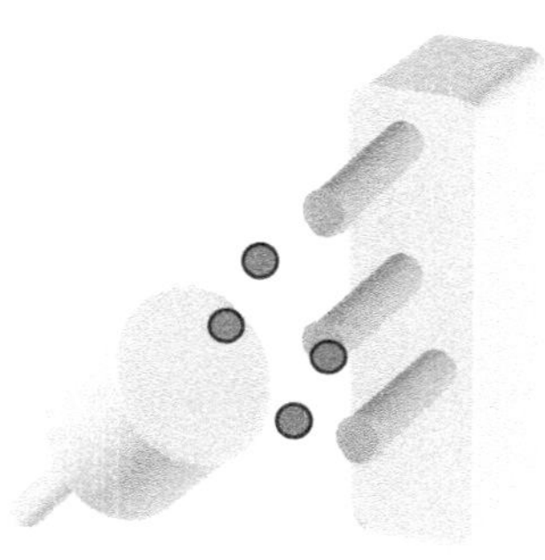

Normal

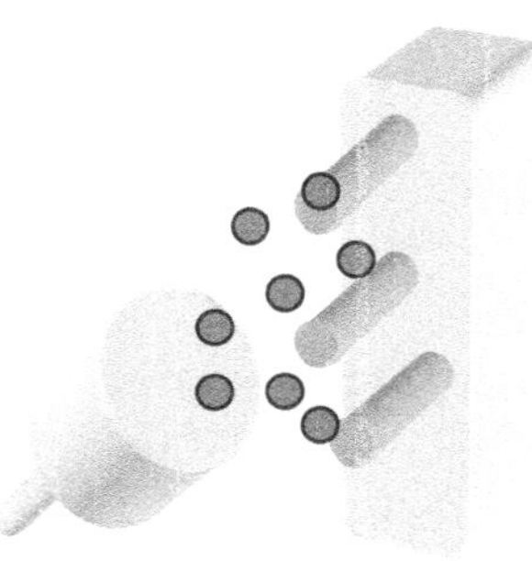

Cocaine
(Increased Dopamine)

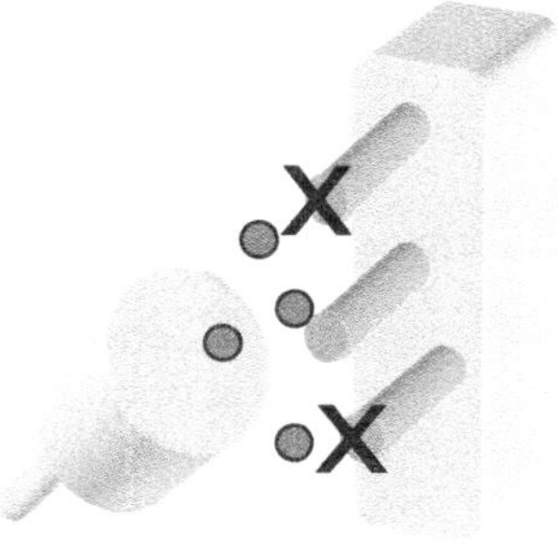

Neuroleptic
(Receptor blockade)

3.

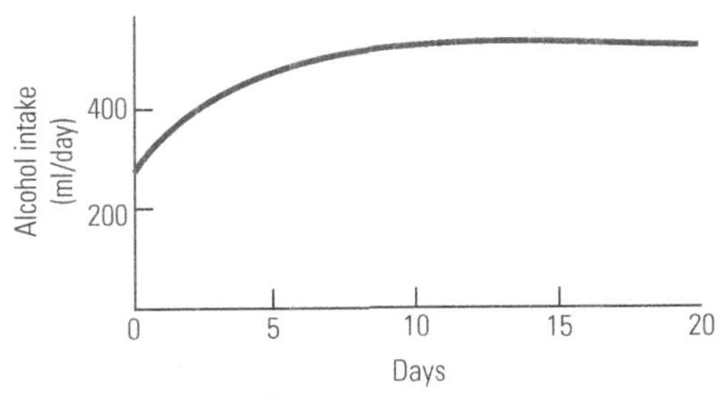
Alcohol intake
(ml/day)
400
200
0
5
10
15
20
Days

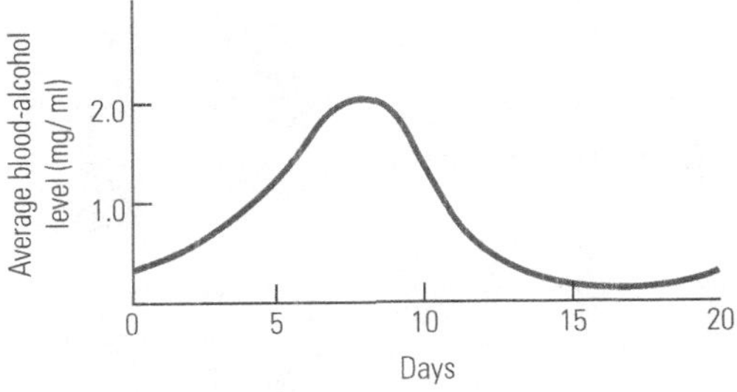
Average blood-alcohol
level (mg/ ml)
2.0
1.0
0
5
10
15
20
Days

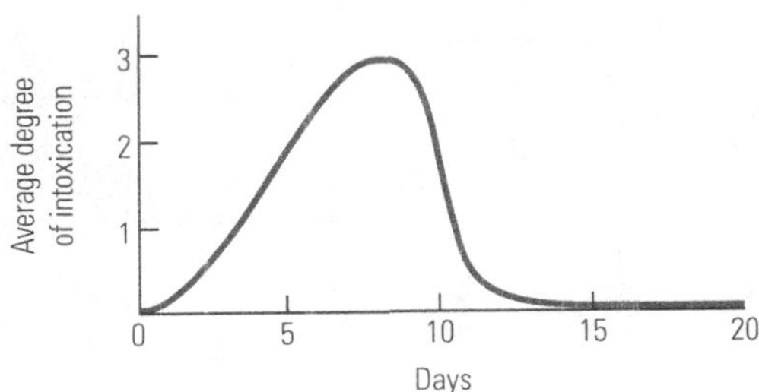
Average degree
of intoxication
3
2
1
0
5
10
15
20
Days

4.

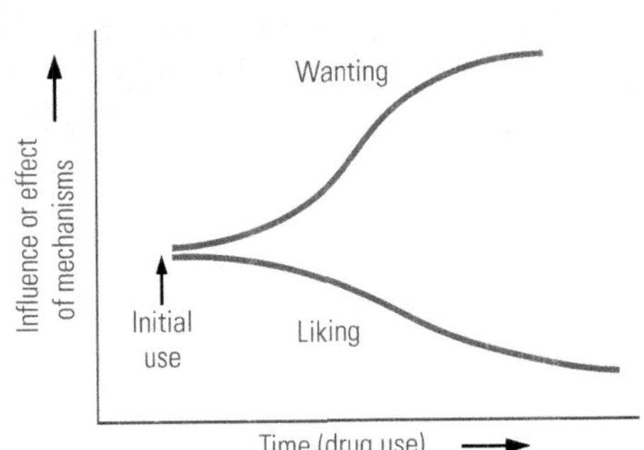
Influence or effect
of mechanisms
Wanting
Initial
use
Liking
Time (drug use)

5.

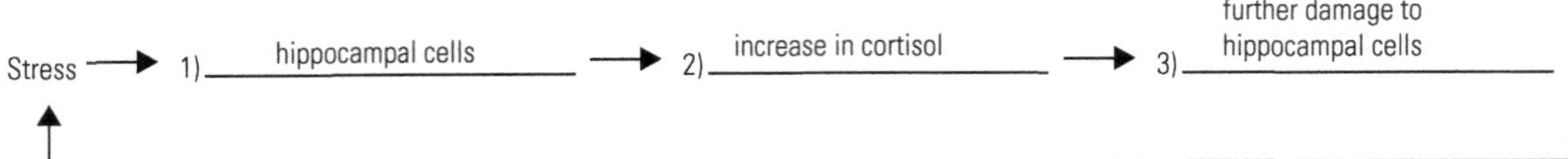

Crossword Puzzle

1 A	N	T	I	D	E	P	R	E	S	2 S	A	N	3 T			4 A	B	U	S	E	R			
N										E			R			M						5 M		
T							6 L			N			A			P						E		
I							S			S			N			H			7 M			S		
A					8 S	E	D	9 A	T	I	V	E	S			E			D			O		
N									S	T			10 P	O	S	T	R	E	M	11 A		L		
X						12 W			S	I			O			A			13 A	C	T	I	V	14 E
I						I			O	Z			R			M				I		M		N
E			15 H			T			C	A			16 T	O	X	I	N			D		B		D
T			O			H			I	T						N						I		O
Y			M			D			A	I		17 S	P	I	D	E	R					C		C
		18 G	E	N	E	R	A	T	I	O	19 N													R
			O			A			I	20 N	A	R	C	O	T	21 I	C			22 B	B	23 B		I
			S			W			V		L					A						R		N
24 G			T			A			25 E	N	D	O	R	P	H	I	N	S	26 C	U	R	A	R	E
A			A			L					X											I		
B			S								27 O	X	I	D	A	S	E					N		
28 A	D	D	I	C	T	I	O	N			N													
			S							29 H	E	R	O	I	N									

Chapter 7: How Do We Study the Brain's Structure and Functions?

Multiple Choice Questions

1. A (p. 217)
2. C (p. 218)
3. C (p. 219)
4. B (p. 220)
5. C (p. 222)
6. C (p. 224)
7. D (p. 225)
8. C (p. 228)
9. A (p. 231)
10. E (p. 232)
11. A (p. 234)
12. C (p. 234)
13. A (p. 236)
14. D (p. 236)
15. E (p. 236)
16. C (p. 237)
17. E (p. 237)
18. D (p. 241)
19. C (p. 242)
20. B (p. 243)

Short-Answer Questions

1. (p. 217)
2. (p. 220)
3. (p. 221)
4. (p. 224)
5. (p. 228)
6. (p. 232)
7. (p. 236)
8. (p. 236)
9. (p. 238)
10. (p. 242)

Matching Questions

1. __C__ Corsi block tapping test
   __B__ Mirror-drawing task
   __D__ Recency memory ttask
   __A__ Matching-to-place learning
   __E__ Paw-reaching behavior
2. __NI__ transcranial magnetic stimulation
   __I__ Deep brain stimulation
   __I__ Electrical lesions
   __NI__ Event-related potential recording (ERP)
   __NI__ Optical tomography
3. __CT__ Uses X-ray
   __fMRI__ Uses a magnetic field
   __PET__ Uses radioactive compounds
   __EEG__ Measures electrical currents
   __NIRS__ Measures light transmission
4. __C__ Kyoto SHR rat
   __D__ Interrupted blood supply to the brain
   __A__ Neurotoxic lesions of dopamine pathways
   __E__ Infant rats deprived of maternal contact
   __B__ Electrolytic lesion of the hippocampus
5. __D__ Whishaw
   __E__ Morris
   __A__ Berger
   __C__ Lashley
   __B__ Penfield
   __F__ Broca

Diagrams:

1.

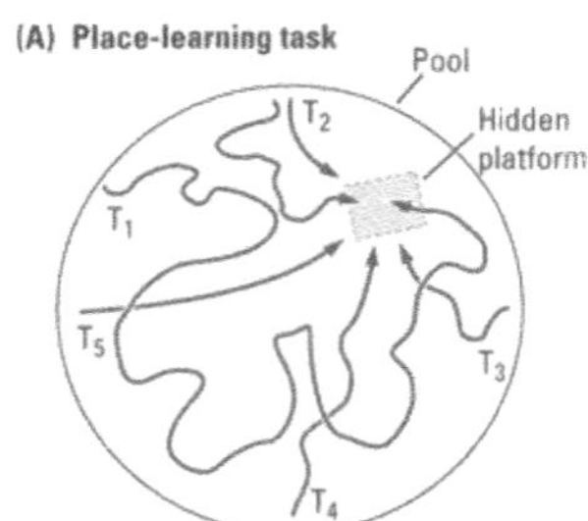

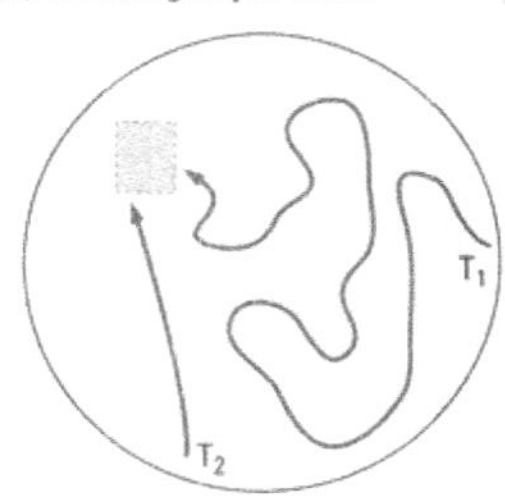

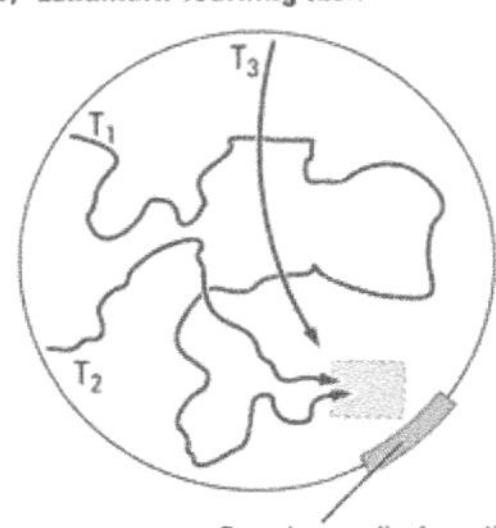

2.

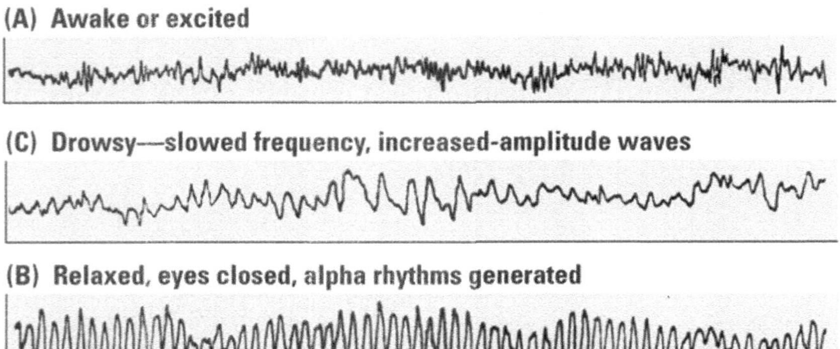

3. Both are viewed from a dorsal perspective.

MRI CT

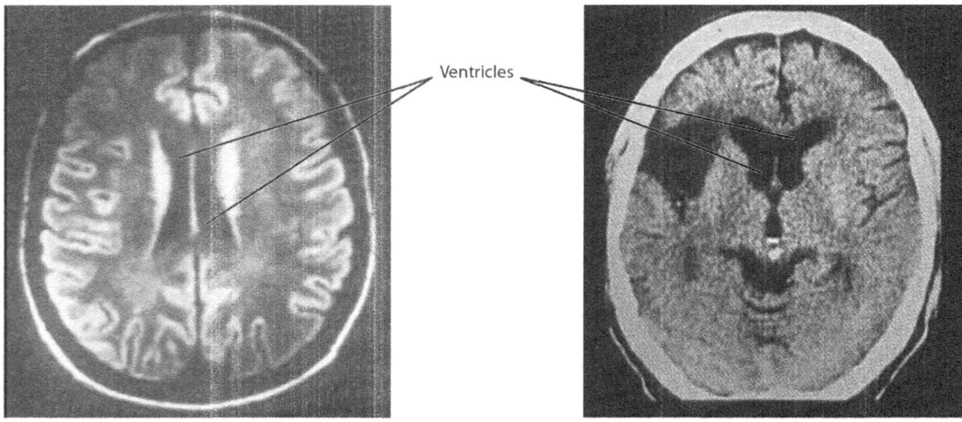

4. A. Transcranial magnetic stimulation
   B. Polygraph or electroencephalograph
   C. Magnetic resonance imaging
   D. Positron emission tomography

5.

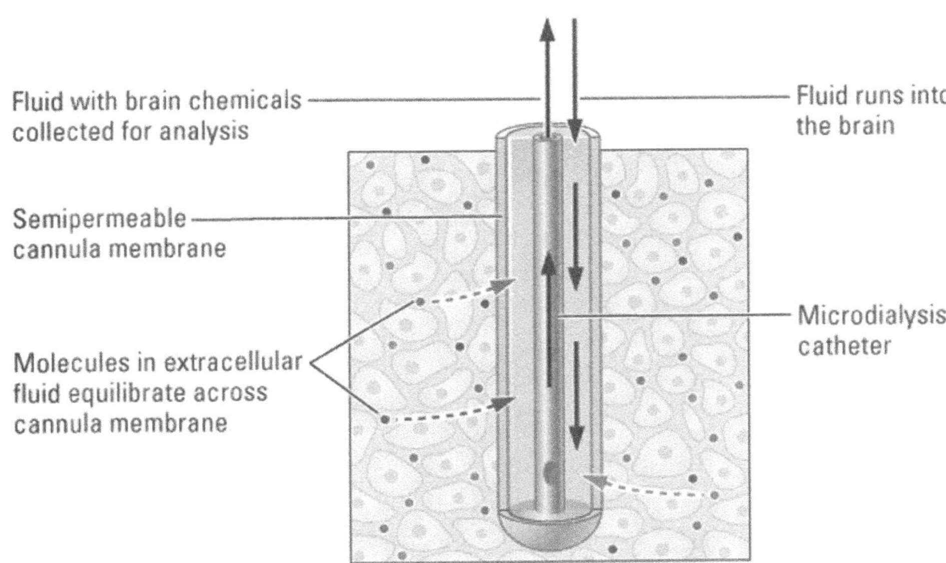

Crossword Puzzle

1 R	E	2 C	O	R	3 D	I	N	G								
		O			B											
		N			S						4 B		5 G			
		C				6 N					R		E			
		O		7 H		I		8 O			A		N			
9 A		R		Y		R		P			I		10 E	11 R	P	
P		D		12 P	O	S	I	T	R	O	N			A		
13 P	L	A	C	E				I						T		
A		N		R				C					14 D	E	E	P
R		15 C	16 T			17 S	C	A	18 N				Y			
19 A	R	E	A		20 E			21 L	E	S	22 I	O	N		23 S	
T			P		V				A		M		24 A	D	H	D
U			P		E				R		A		M		R	
S			I		N						G		I			
	25 K	I	N	E	T	I	C				I		26 C	O	M	27 T
			G								N					M
									28 E	E	G					S

Chapter 8: How Does the Nervous System Develop and Adapt?

Multiple-Choice Questions

1. D (p. 247)
2. D (p. 248)
3. E (p. 250)
4. B (p. 250)
5. C (p. 252)
6. D (p. 253)
7. B (p. 253)
8. D (p. 254)
9. B (p. 255)
10. E (p. 257)
11. D (p. 260)
12. A (p. 263)
13. A (p. 264)
14. E (p. 268)
15. D (p. 269)
16. A (p. 273)
17. E (p. 274)
18. D (p. 275)
19. A (p. 277)
20. B (p. 279)

Short-Answer Questions

1. (p. 248)
2. (p. 253)
3. (p. 257)
4. (p. 258)
5. (p. 264)
6. (p. 267)
7. (p. 269)
8. (p. 270)
9. (p. 277)
10. (p. 279)

Matching Questions

1. _4_ Progenitor cells
   _1_ Neural groove
   _3_ Neural stem cells
   _5_ Neuroblasts and glioblasts
   _2_ Neural tube

2. _A_ Stimulates production of progenitor cells
   _D_ Guide growth cones to the cell
   _C_ Guide cell migration from neural tube
   _B_ Stimulates production of neuro-blasts
   _E_ Provide adhesive surface for guiding cells

3. _C_ Axonal projections
   _B_ Dendritic branching
   _D_ Programmed cell death
   _A_ Cell proliferation
   _E_ Process of eliminating neurons

4. _D_ Formal operational
   _C_ Sensorimotor
   _A_ Concrete operational
   _B_ Preoperational

5. _A_ Lazy eye syndrome
   _C_ Serious motor problems
   _B_ Abnormal intellect and social behaviors
   _E_ Genetic adnormality
   _D_ Fatal soon after birth

Diagrams

1.

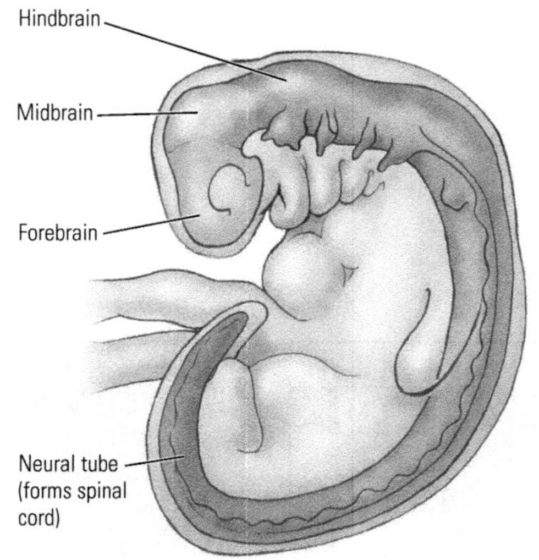

2.

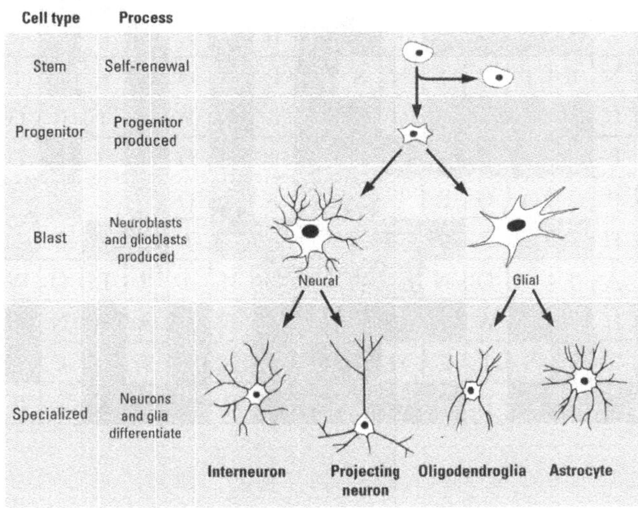

3.

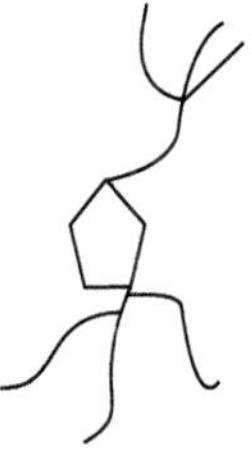

Neuron B

Neuron C

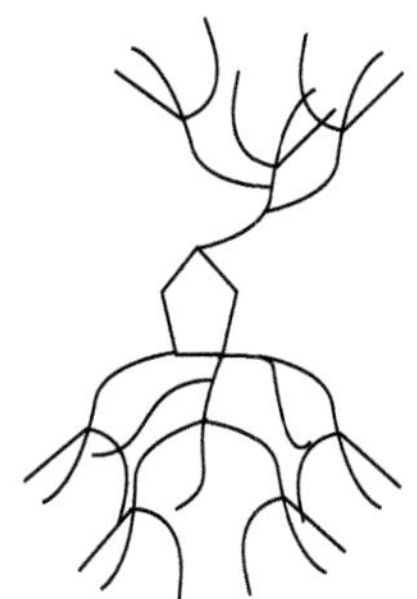

4.

5. C

Crossword Puzzle

1 N	E	U	R	O	2 G	E	N	E	S	3 I	S									4 T				
E					R					M										R				
O					O					5 P	K	U					6 B	I	F	I	7 D	A		
					W		8 C			R						9 G					E		10 S	
							E			I						L					N		P	
					11 A		L			N			12 R		13 F	I	L	O	14 P	O	D	I	U	M
15 D		16 P	H	E	N	Y	L	K	E	T	O	N	U	R	I	A			E		R		R	
O		L			E					I			B		L		17 F		R		I		T	
18 W	R	A	P		19 N	E	T	R	I	N	S		E		O		E		I		T		S	
N		S			C					G			L		P		20 T	R	O	P	I	21 C		
S		22 T	U	B	E								L		O		A		D		23 C	A	M	
		I			24 P	I	A	25 G	E	T			A		D		L					M		
		C			H			R							I			26 C				S		
	27 M	I	G	R	A	T	I	O	N					28 H	A	29 R	L	O	W					
		T			L			W						E		A		N						
		Y			Y		30 S	T	E	M				B		T		E						
								H						B				31 S	P	I	N	A		

Chapter 9: How Do We Sense, Perceive, and See the World?

Multiple-Choice Questions

1. C (p. 282)
2. D (p. 286)
3. D (p. 288)
4. D (p. 288)
5. E (p. 290)
6. E (p. 291)
7. D (p. 293)
8. D (p. 294)
9. A (p. 296)
10. C (p. 298)
11. D (p. 302)
12. E (p. 304)
13. B (p. 305)
14. D (p. 306)
15. C (p. 309)
16. D (p. 311)
17. C (p. 313)
18. A (p. 314)
19. B (p. 314)
20. E (p. 316)

Short-Answer Questions

1. (p. 283)
2. (p. 288)
3. (p. 288)
4. (p. 292)
5. (p. 302)
6. (p. 303)
7. (p. 306)
8. (p. 310)
9. (p. 310)
10. (p. 314)

Matching Questions

1. _D_ Contains photoreceptors
   _B_ Directs image onto the fovea
   _C_ Controls amount of light entering
   _E_ Contains blood vessels entering eye
   _A_ Outer covering of eye

2. _R_ used primarily for night vision
   _C_ used for color vision
   _C_ highest density found in the fovea
   _C_ used for acute vision
   _R_ long and slender in shape
   _R_ most of the receptors in the eye are this type

3. _M_ receive input primarily from rods
   _M_ found in the periphery of the retina
   _P_ are sensitive to color
   _P_ are the smaller of the two ganglion cells
   _M_ are more sensitive to light

4. _E_ Deficit in reaching using visual guidance
   _A_ Usually compensated for by nystagmus
   _B_ Caused by complete lesion of one optic tract
   _C_ More common in males than females
   _D_ Inability to recognize objects

5. _2_ Optic chiasm
   _4_ Occipital lobe
   _3_ Lateral geniculate nucleus
   _1_ Fovea
   _5_ Temporal lobe

Diagrams

1.

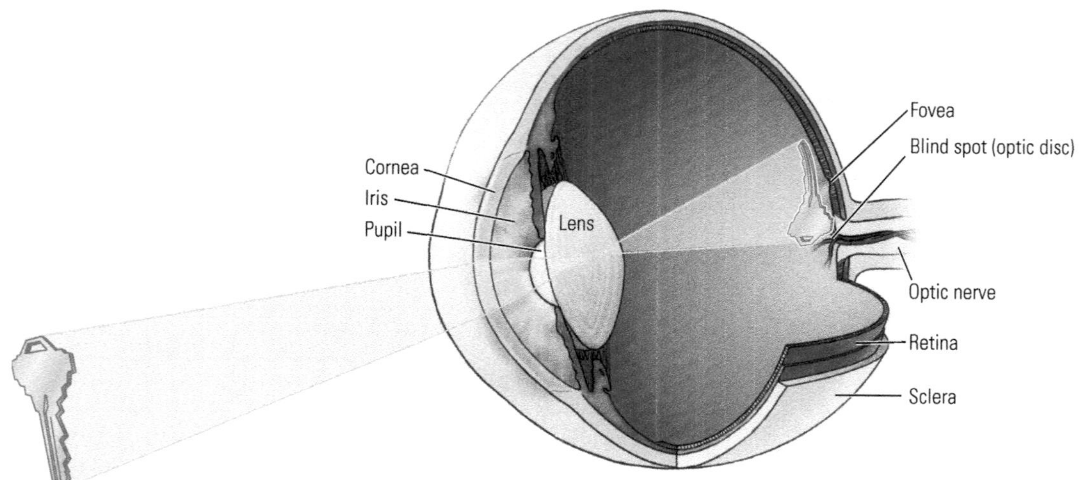

2.

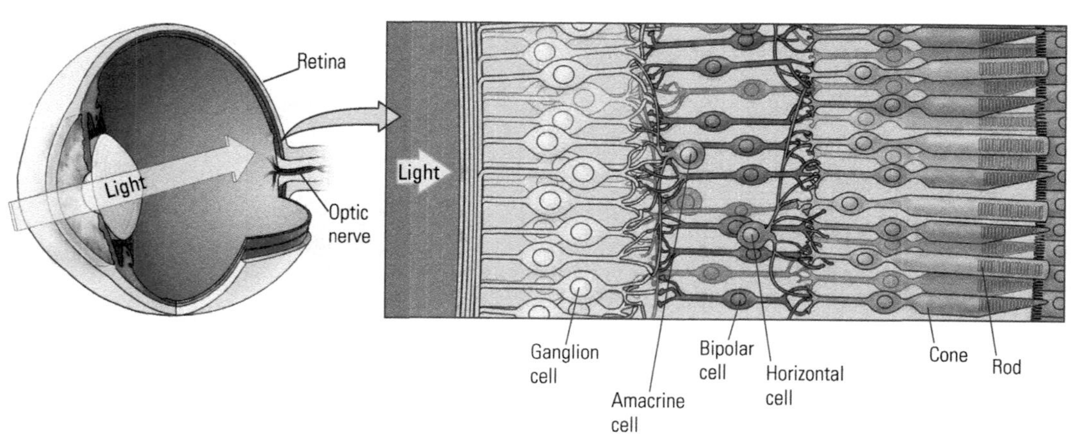

3.

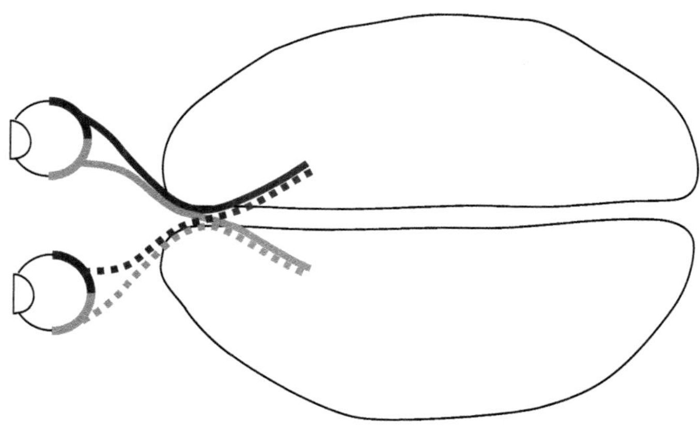

4. A = Green

   B = Red

5.

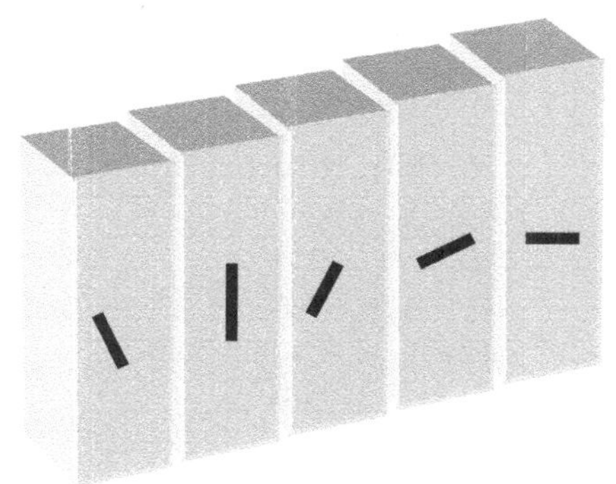

Crossword Puzzle

1 C	O	R	N	E	2 A						3 D							4 S	5 P	O	T			
O					6 T	R	I	C	H	R	O	M	A	T	I	7 C			R					
M					A						R					8 O	P	T	I	C			9 H	
P					X		10 B	L	O	B	S					L			M		11 L		E	
12 L	G	N			I						A					O		13 L	A	T	E	R	A	L
E				14 P	A	R	V	O	C	15 E	L	L	U	L	A	R			R		N		R	
X										X								16 E	Y	E	S			
								17 I	N	T	E	R	18 B	L	19 O	B	20 S							
					21 P					R			L		N		22 T	H	E	O	R	I	E	23 S
			24 T		U			25 A	M	A	C	R	I	N	E		R							I
			H		P					S			N			26 B	I	P	O	L	A	27 R		M
		28 G	E	N	I	C	U	L	A	T	E		D				A					O		P
			O		L					R		29 I					T		30 D			D		L
31 T	E	A	R		32 S	U	P	E	R	I	O	R				33 R	E	T	I	N	A	S		E
			Y							A		I				G			S					
										T		S			34 E	Y	E		35 C	O	R	T	E	X
36 H	Y	P	E	R	C	O	M	P	L	E	X					B								

Chapter 10: How Do We Hear, Speak, and Make Music?

Multiple-Choice Questions

1. C (p. 320)
2. A (p. 321)
3. D (p. 322)
4. C (p. 322)
5. D (p. 323)
6. E (p. 326)
7. E (p. 328)
8. B (p. 329)
9. A (p. 330)
10. E (p. 331)
11. B (p. 333)
12. C (p. 334)
13. C (p. 336)
14. A (p. 336)
15. B (p. 338)
16. E (p. 339)
17. C (p. 341)
18. A (p. 345)
19. E (p. 348)
20. E (p. 350)

Short-Answer Questions

1. (p. 324)
2. (p. 329)
3. (p. 335)
4. (p. 336)
5. (p. 338)
6. (p. 339)
7. (p. 339)
8. (p. 340)
9. (p. 348)
10. (p. 351)

Matching Questions

1. _B_ Measured in decibels
   _E_ Primarily right hemisphere
   _A_ Measured in hertz
   _C_ Time difference between ears
   _D_ Posterior temporal lobe

2. _2_ Ganglion cells
   _5_ Inferior colliculus
   _1_ Bipolar cells
   _3_ Cochlear nucleus
   _4_ Superior olive

3.

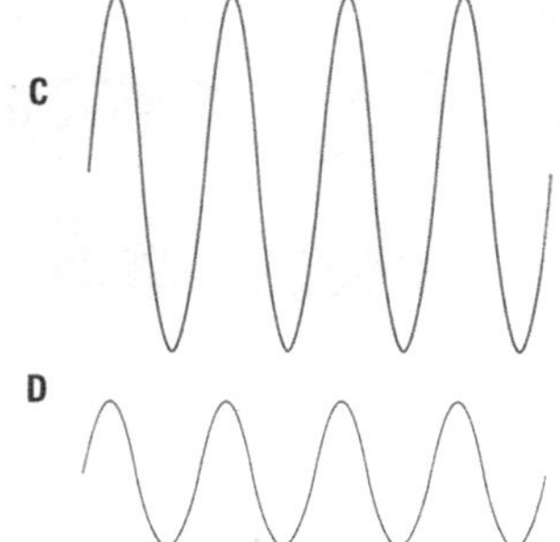

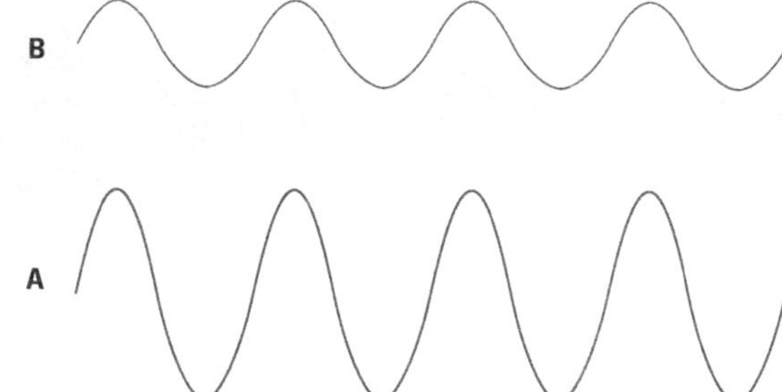

4. A Cortex
   D Hindbrain
   E Cochlea
   B Thalamus
   C Midbrain

5. D From the right
   B From ahead
   C From behind
   A From the left

Diagrams

1.

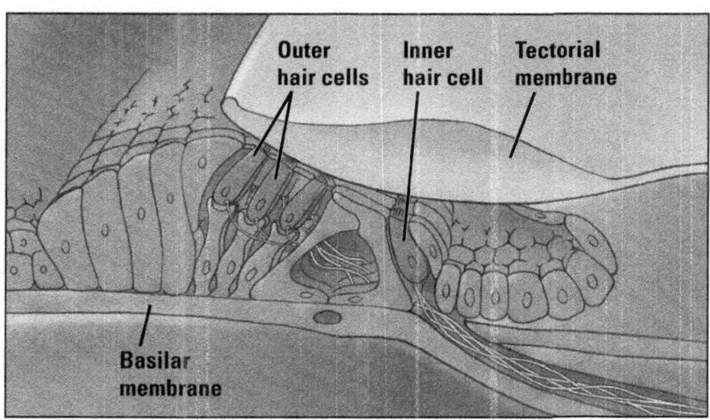

2. A. Tuba
   B. Trumpet
   C. Flute
3. B
4.

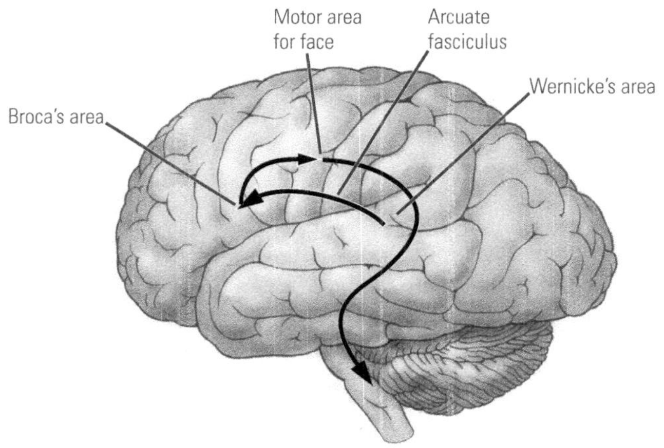

5.

C. A. B. D.

0 20 40 60 80 100 120 140 160 180 200

Crossword Puzzle

1 F	R	2 E	Q	U	3 E	N	C	I	E	S			4 D				5 A	6 G	E				
O		C			X								R					A					
R		H			T			7 N	U	8 C	L	E	U	S				N			9 P		
K		O			E		10 P			O			M		11 B			G			E		
S		L			R		O			C					12 I	M	P	L	A	N	T	S	
	13 L	O	U	D	N	E	S	S		14 H	A	M	M	E	R			I					
		C			A		I			L					D			O		15 D	16 E	A	F
		17 A	M	P	L	I	T	U	18 D	E								N			M		
		T					R		19 E	A	R		20 N		21 S						I		
		I		22 H			O		C				O		O						S		
23 B	R	O	C	A	S		N		I				I		U		24 A	25 P	H	A	S	I	A
R		N		I					B				26 S	27 O	N	G		E			I		
O				R				28 P	E	N	F	I	E	L	D		29 P	R	O	S	O	D	Y
30 C	R	O	W		31 A	N	V	I	L					I				F			N		
A								N						V				E					
		32 M	E	M	B	R	A	N	E	S			33 W	E	R	N	I	C	K	E			
								A										T					

Chapter 11: How Does the Nervous System Respond to Stimulation and Produce Movement?

Multiple-Choice Questions

1. E (p. 355)
2. B (p. 355)
3. B (p. 358)
4. E (p. 360)
5. C (p. 363)
6. A (p. 362)
7. A (p. 363)
8. A (p. 365)
9. D (p. 365)
10. C (p. 369)
11. A (p. 369)
12. E (p. 368)
13. B (p. 372)
14. C (p. 375)
15. D (p. 378)
16. E (p. 381)
17. C (p. 384)
18. E (p. 386)
19. B (p. 388)
20. B (p. 390)

Short-Answer Questions

1. (p. 355)
2. (p. 360)
3. (p. 365)
4. (p. 373)
5. (p. 374)
6. (p. 375)
7. (p. 379)
8. (p. 388)
9. (p. 390)
10. (p. 394)

Matching Questions

1. _A_ "Volume control" for movement
   _B_ Planning movement
   _D_ Monosynaptic reflexes
   _C_ Executing movement
   _E_ Regulating posture

2. _B_ Somatosensory cortex
   _A_ Brainstem
   _E_ Spinal cord
   _D_ Cerebellum
   _C_ Basal ganglia

3. _B_ Utilizes free nerve endings that release chemicals
   _C_ Encapsulated nerve endings monitoring tendons
   _B_ Utilizes small unmyelinated fiber
   _A_ Responds best to pressure
   _C_ Processes information on body position

4. _2_ Dorsal column nuclei
   _4_ Ventrolateral thalamus
   _1_ Muscle stretch receptors
   _3_ Medial lemniscus
   _5_ Somatosensory cortex

5. _E_ May send excitatory or inhibitory signals
   _B_ Senses changes in 3 planes of different orientations
   _D_ Small crystals of calcium carbonate
   _C_ Fluid filling the semicircular canals
   _A_ Consists of the ultricle and saccule

Diagrams

1.

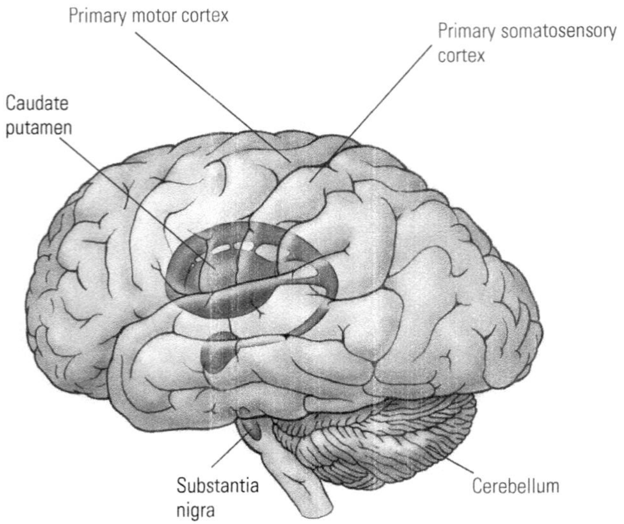
Primary motor cortex
Primary somatosensory cortex
Caudate putamen
Substantia nigra
Cerebellum

2.

3.

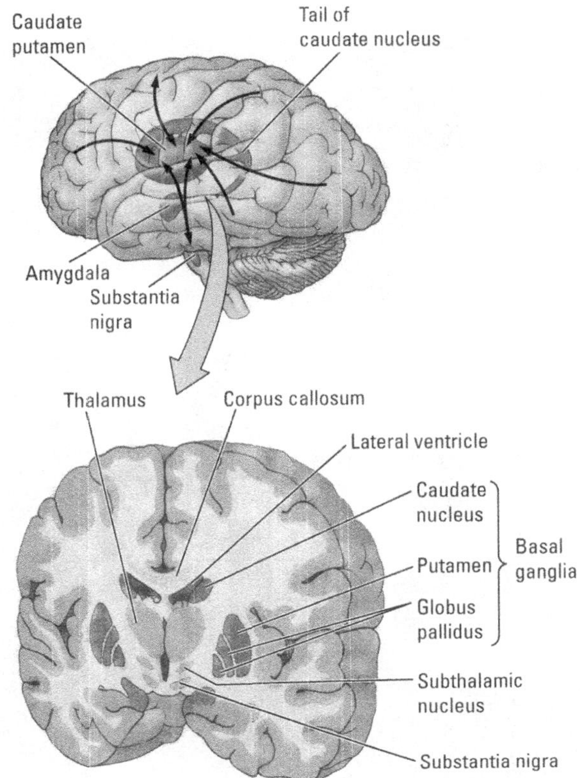
Caudate
putamen
Tail of
caudate nucleus
Amygdala
Substantia
nigra
Thalamus
Corpus callosum
Lateral ventricle
Caudate
nucleus
Putamen
Globus
pallidus
Basal
ganglia
Subthalamic
nucleus
Substantia nigra

4.

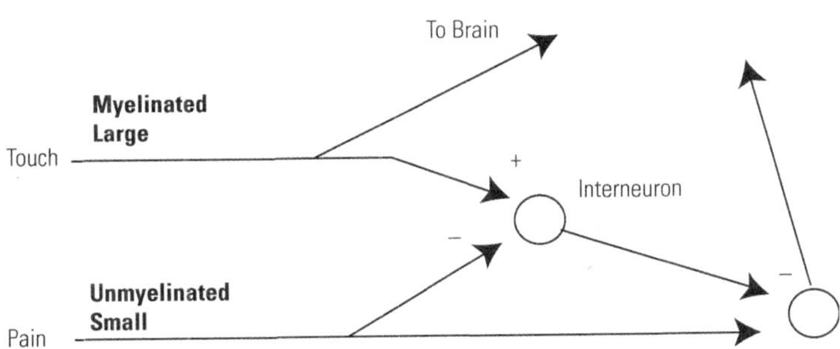
To Brain
Myelinated
Large
Touch
+
Interneuron
–
Unmyelinated
Small
Pain
–

5.

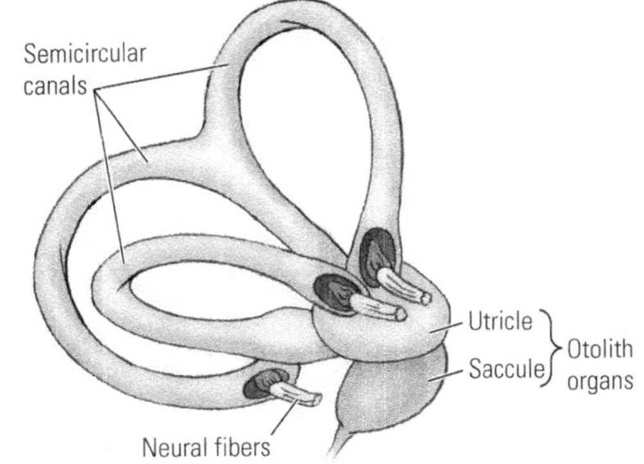
Semicircular
canals
Utricle
Saccule
Otolith
organs
Neural fibers

Crossword Puzzle

1. VENTROLATERAL / VESTIBULAR
2. TOURETTES
3. PAIN
4. MOTOR
5. PYRAMIDAL
6. KINETIC
7. ROOT
8. MEDIAL
9. DISSOLUTION
10. CORTICOSPINAL
11. CORD / CANALS
12. ENDOLYMPH
13. PENFIELD
14. SYNERGIES
15. LASHLEY
16. OTOCONIA
17. HORNS
18. DORSAL
19. SYSTEM
20. HAPSIS
21. PALSY
22. SPINAL
23. UTRICLE
24. SKIN
25. GANGLIA
26. OTOLITH

Chapter 12: What Causes Emotional and Motivated Behavior?

Multiple-Choice Questions

1. A (p. 400)
2. A (p. 403)
3. B (p. 408)
4. E (p. 411)
5. D (p. 411)
6. A (p. 413)
7. B (p. 416)
8. D (p. 422)
9. D (p. 423)
10. A (p. 425)
11. C (p. 425)
12. D (p. 426)
13. B (p. 426)
14. B (p. 430)
15. C (p. 430)
16. E (p. 432)
17. C (p. 433)
18. B (p. 434)
19. C (p. 436)
20. A (p. 438)

Short-Answer Questions

1. (p. 406)
2. (p. 409)
3. (p. 411)
4. (p. 422)
5. (p. 423)
6. (p. 424)
7. (p. 426)
8. (p. 433)
9. (p. 436)
10. (p. 438)

Matching Questions

1. D Lack of prosody
   A Reduced fear response
   C Hyperphagia
   E Disruption of male sexual behavior
   B Aphagia behavior

2. C Treated with anxiolytic drugs
   A Reduced dopamine from frontal lobe
   B Disruption of serotonin and noradrenaline
   C Overactivity of $GABA_A$ receptors
   B Affects nearly 10 percent of the population

3. C Satiety
   E Fear response
   A Hypovolemic thirst
   B Osmotic thirst
   D Sexual receptivity

4. A Persistent and unrealistic worries
   C Involve a clearly dreaded object or situation
   B Recurrent attacks of intense terror
   B Often leads to agoraphobia
   C Most common type of anxiety disorder

5. O Testosterone producing sexual dimorphism of preoptic area
   A Testosterone producing male sexual behavior
   A Estrogen and progesterone producing lordosis
   O Congenital adrenal hyperplasia causing an enlarged clitoris
   A Estrous cycle hormones increasing dendritic branching

Diagrams

1.

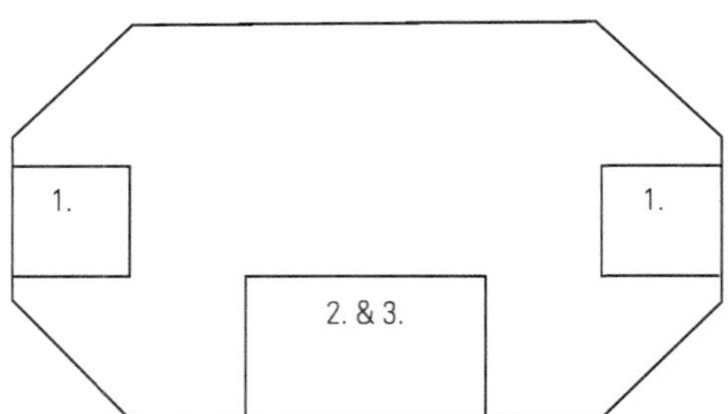

2.

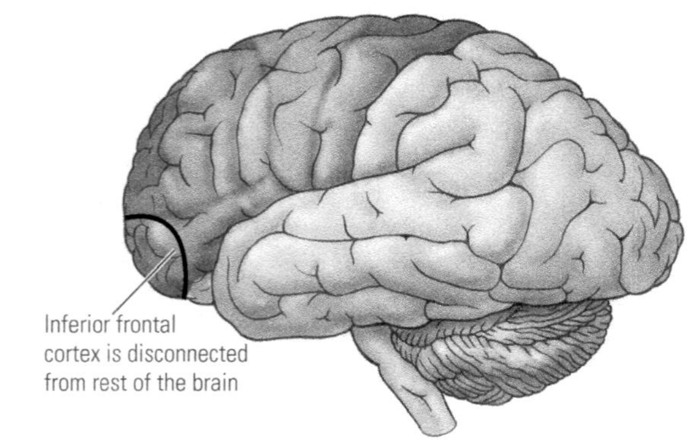

3. All cells project to posterior pituitary

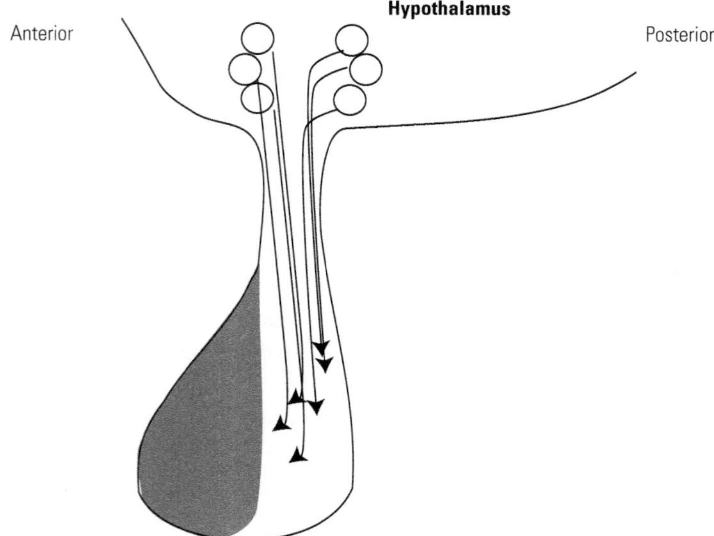

4. Add extracellular salt.

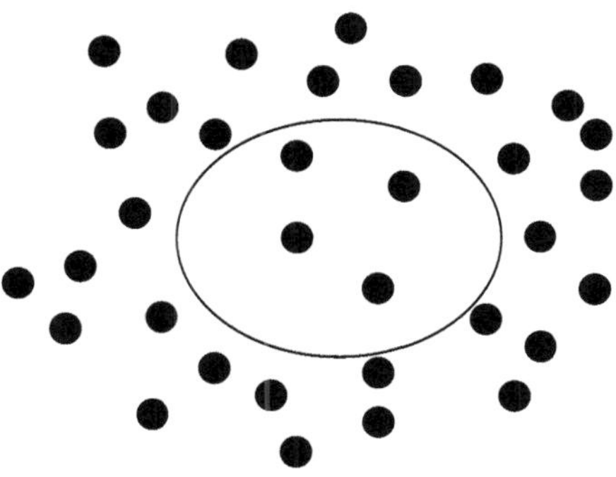

5. Neurons A, C, & D would be masculinized.

Crossword Puzzle

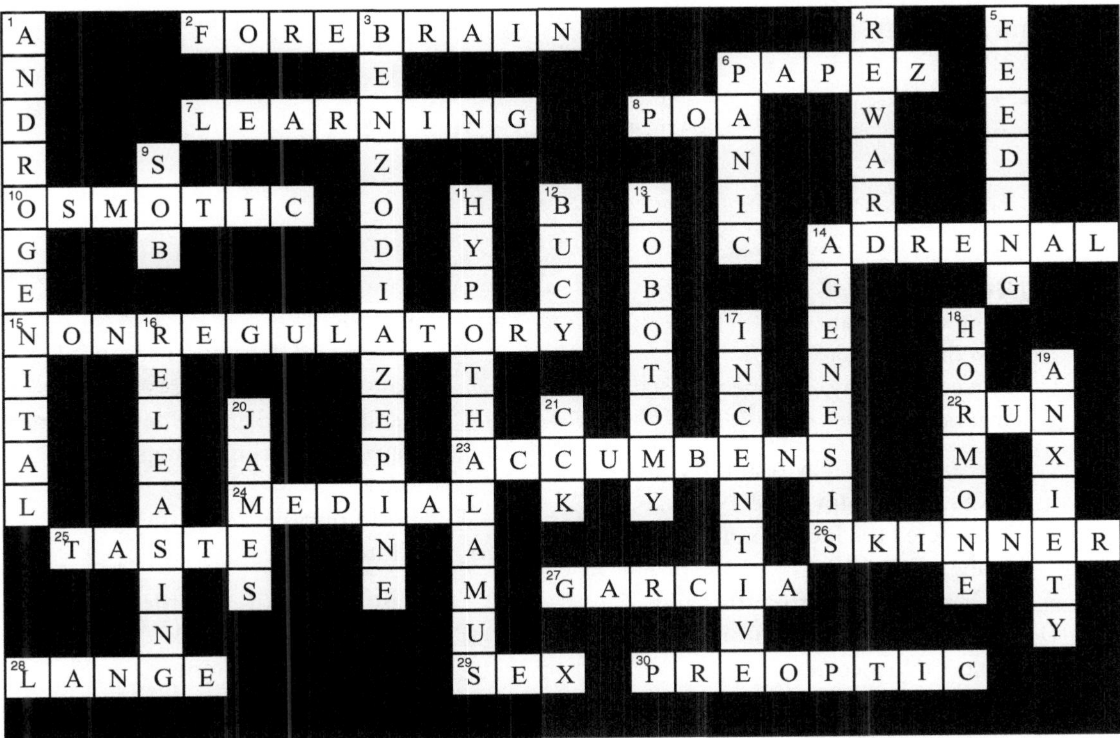

Chapter 13: Why Do We Sleep and Dream?

Multiple-Choice Questions

1. A (p. 447)
2. D (p. 448)
3. C (p. 449)
4. B (p. 452)
5. E (p. 452)
6. A (p. 455)
7. D (p. 459)
8. D (p. 460)
9. A (p. 462)
10. E (p. 463)
11. E (p. 465)
12. B (p. 465)
13. E (p. 468)
14. A (p. 469)
15. D (p. 471)
16. C (p. 472)
17. B (p. 473)
18. E (p. 473)
19. E (p. 477)
20. A (p. 478)

Short-Answer Questions

1. (p. 451)
2. (p. 452)
3. (p. 454)
4. (p. 456)
5. (p. 464)
6. (p. 465)
7. (p. 467)
8. (p. 469)
9. (p. 475)
10. (p. 476)

Matching Questions

1. A Around one year
   D Less than one day
   C Between one day and one year
   B Around one day

2. C Damage to this may result in coma
   E PGO spikes originate here
   A Considered the location of the main biological clock
   B EEGs are recorded from this area
   D Contains serotonin neurons that project to neocortex

3. NREM Delta rhythms
   REM Paralysis
   NREM Night terrors
   NREM Sleepwalking
   REM Loss of temperature-regulatory mechanism

4. A More common in people who are overweight
   B Anxiety and depression account for about 35 percent of cases
   C Results from tolerance development
   D May occur during a cataplexy attack
   E NREM disorder seen especially in children

5. D Seasonal affective disorder
   B Cataplexy
   C Restless Legs Syndrome
   A REM without atonia
   B Narcolepsy

Diagrams

1.

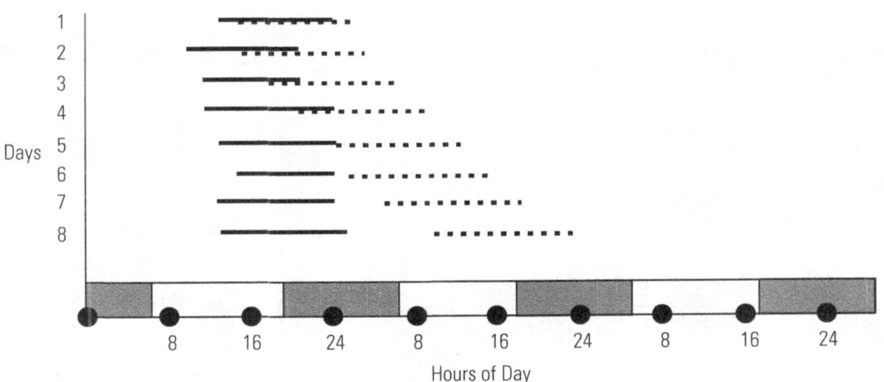

2. Traveler C.

3.

Stage 1:

Stage 4:

Awake and alert:

4.

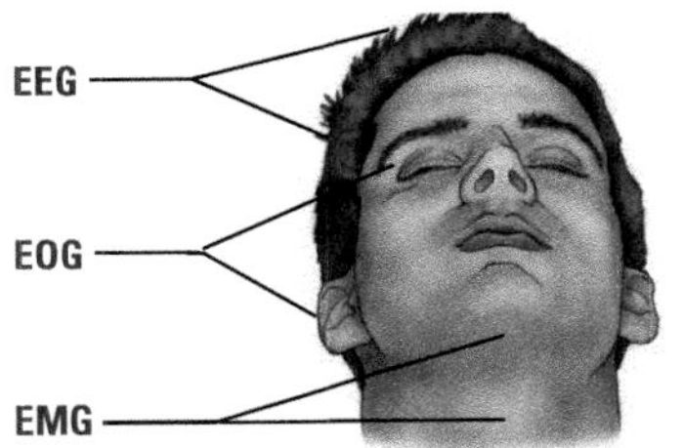

5.

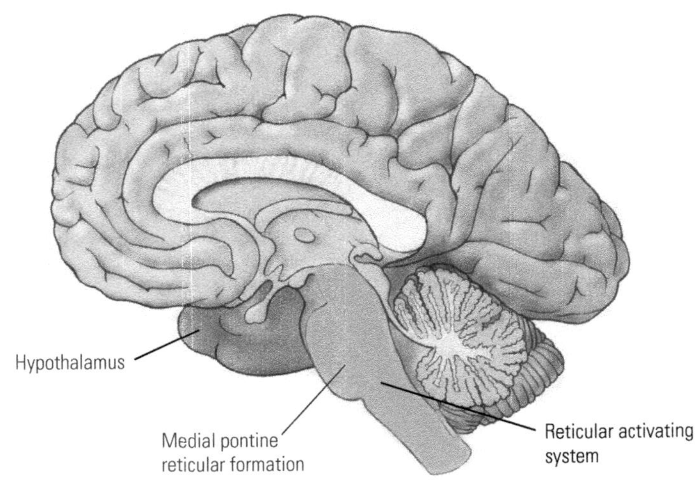

Crossword Puzzle

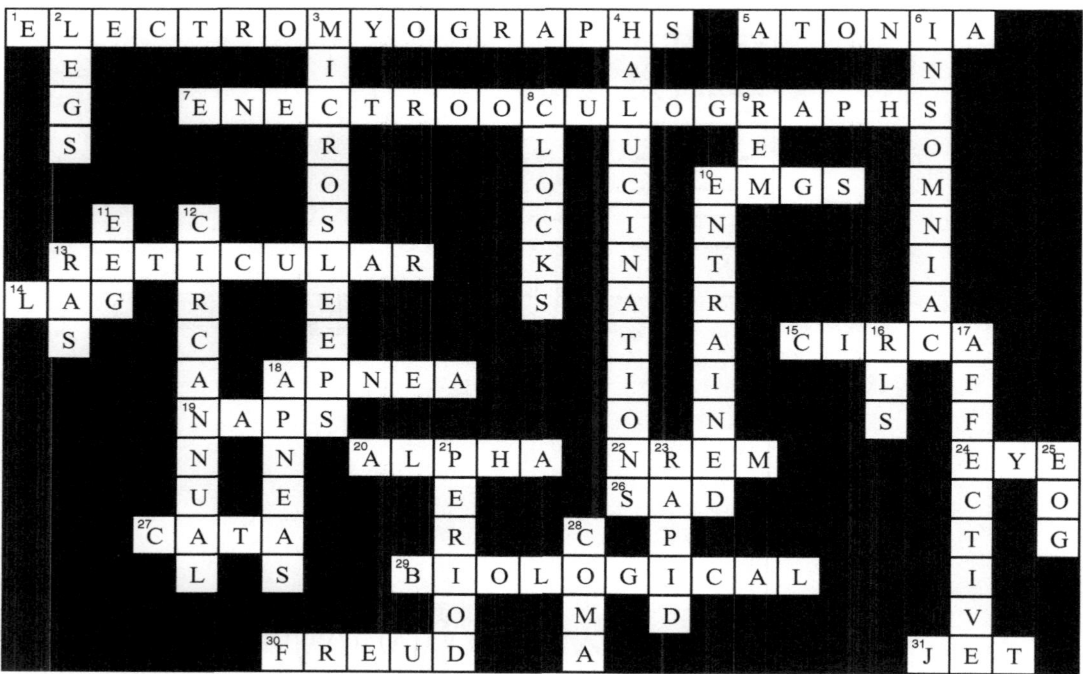

Chapter 14: How Do We Learn and Remember?

Multiple-Choice Questions

1. D (p. 483)
2. B (p. 484)
3. C (p. 485)
4. B (p. 486)
5. D (p. 487)
6. A (p. 488)
7. D (p. 491)
8. C (p. 491)
9. E (p. 494)
10. B (p. 497)
11. C (p. 500)
12. C (p. 500)
13. D (p. 503)
14. C (p. 506)
15. C (p. 510)
16. A (p. 512)
17. A (p. 513)
18. E (p. 514)
19. D (p. 518)
20. E (p. 520)

Short-Answer Questions

1. (p. 483)
2. (p. 486)
3. (p. 489)
4. (p. 490)
5. (p. 491)
6. (p. 493)
7. (p. 497)
8. (p. 510)
9. (p. 517)
10. (p. 518)

Matching Questions

1. OC A dog waiting near the table to be fed food scraps
   CC Ducking your head when you hear a loud noise
   CC The feeling of hunger when you smell pizza
   OC Studying late into the night before an exam
   OC Holding the door open for someone entering a building behind you

2. IM Riding a bicycle
   EM Childhood memories
   EM Declarative memory
   IM Procedural memory
   EM Top-down processing
   IM Bottom-up processing

3. E Visual object memory
   B Short-term memory
   C Emotional memory
   D Implicit memory
   A Object location

4. C Degeneration of frontal lobe
   A Loss of implicit memory
   B Degeneration of entorhinal cortex
   C Caused by a vitamin deficiency
   A Degeneration of the basal ganglia

5. B Recommends use of speech therapy
   A Undamaged regions will assume control of behavior
   B Recommends use of nerve growth factor
   C Recommends use of epidermal growth factor (EGF)
   C Has been used in many Parkinson's patients

Diagrams

1.

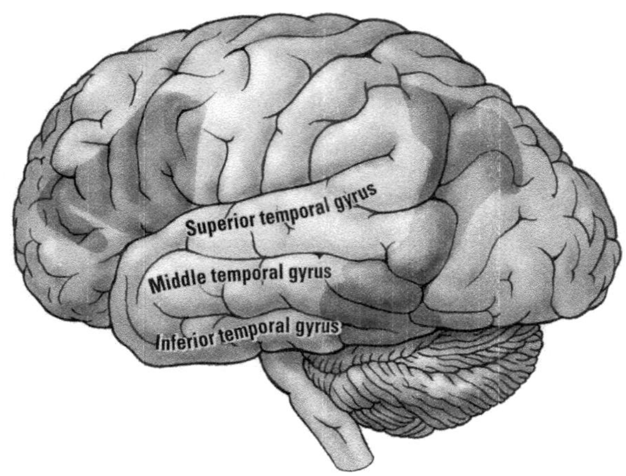

2.

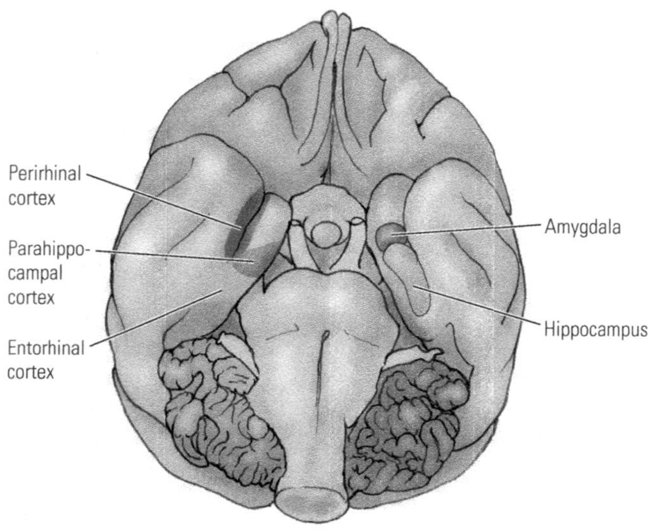

3.

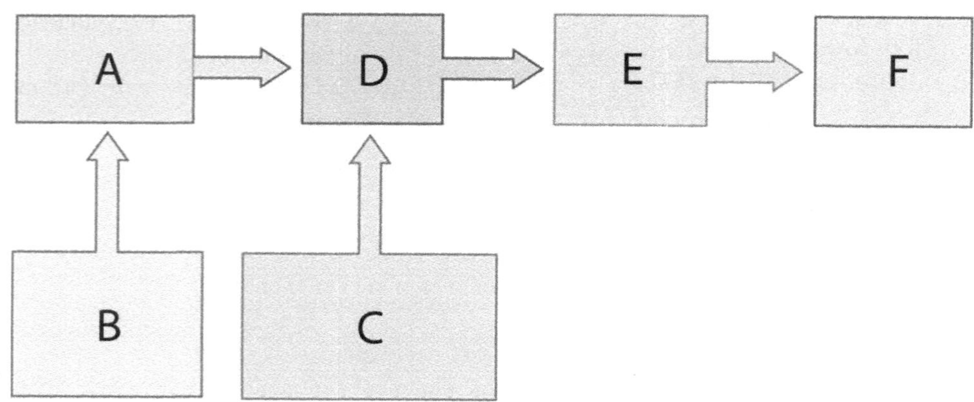

4.

5.

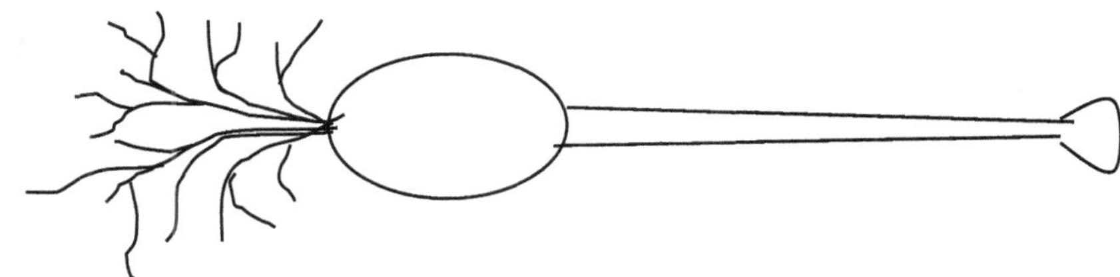

Crossword Puzzle

1 P	A	V	L	O	V																			
A							2 C	3 O	N	D	I	T	I	O	N	I	N	G						
V								P											4 B	5 F				
L		6 S		7 W	O	L	V	E	S		8 M	O	N	K	9 E	Y				A				10 B
O		E						R		11 F					X					12 C	A	J	A	L
V		N						A		E			13 I	M	P	L	I	C	I	T				I
14 I	N	S	T	R	U	M	E	N	15 T	A	16 L				L					O				N
A		I						17 T	H	R	E	E			I		18 E			R				K
N		T							O		G		19 S		C		Y				20 C			
		I							R		G		21 E	P	I	D	E	R	M	A	L			
22 K		Z		23 R	E	S	P	O	N	D	E	N	T		T						A			
E		A							D		D								24 K		S			
G		T							I				25 N		26 F	A	C	T	O	R	S			
27 S	K	I	N	N	E	R			K				E								I			
		O						28 D	E	29 C	L	A	R	A	T	I	V	E			C			
30 S	Y	N	D	R	O	M	E			A			V								A			
										T			31 E	M	O	T	I	O	N	A	L			

Chapter 15: How Does the Brain Think?

Multiple-Choice Questions

1. E (p. 525)
2. B (p. 526)
3. C (p. 526)
4. D (p. 530)
5. E (p. 533)
6. B (p. 536)
7. C (p. 536)
8. E (p. 538)
9. B (p. 544)
10. E (p. 544)
11. A (p. 546)
12. E (p. 547)
13. C (p. 552)
14. B (p. 554)
15. D (p. 554)
16. C (p. 556)
17. A (p. 557)
18. B (p. 559)
19. E (p. 560)
20. E (p. 562)

Short-Answer Questions

1. (p. 525)
2. (p. 526)
3. (p. 535)
4. (p. 536)
5. (p. 538)
6. (p. 544)
7. (p. 547)
8. (p. 556)
9. (p. 559)
10. (p. 560)

Matching Questions

1. B Map reading
   A Language production
   C Temporal planning
   B Music appreciation
   A Language comprehension
2. DT Disrupted particularly by frontal-lobe damage
   CT Measured with traditional intelligence tests
   CT Solving arithmetic problems or defining words
   DT Used to generate numerous answers for a single question
   CT Disruption is often associated with apraxia or aphasia
3. E Ignoring sensory information, usually on one side of the body
   B Inability to make voluntary movements
   C Causes difficulty with the Wisconsin Card Sorting task
   A Caused by damage to the corpus callosum
   D Joining sensory experiences across sensory modalities
4. B Pattern or structure of word order in a phrase
   E The temporal organization of behavior
   A Idea resulting from a set of impressions
   D A form of neglect
   C Term for a wide range of mental abilities
5. NV Left hand
   IV Right hand
   NV Left visual field
   IV Right visual field
   IV Center of visual field

Diagrams

1.

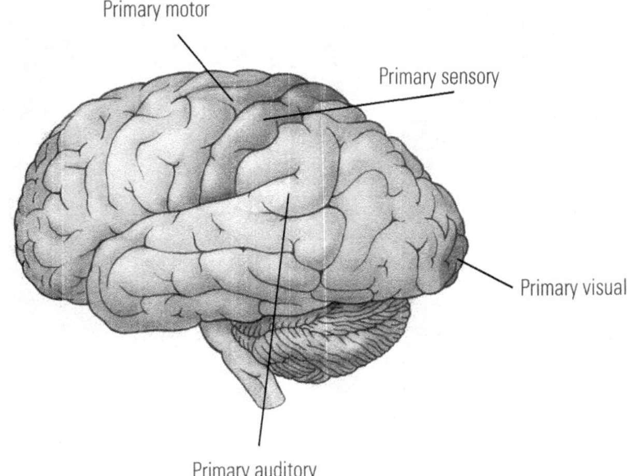

2.

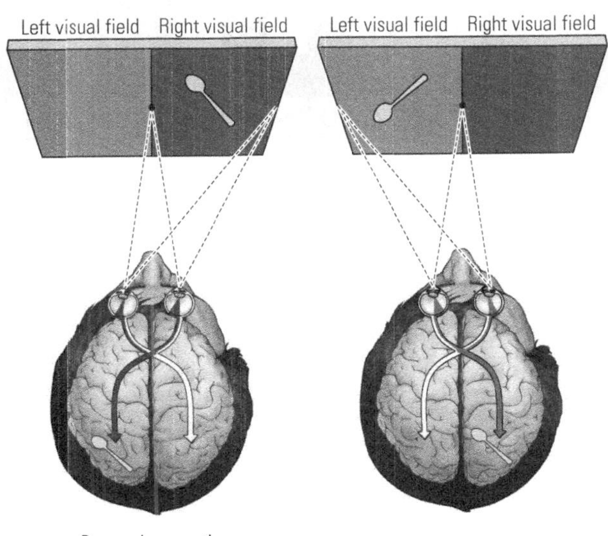

3. B.
4. B.
5.

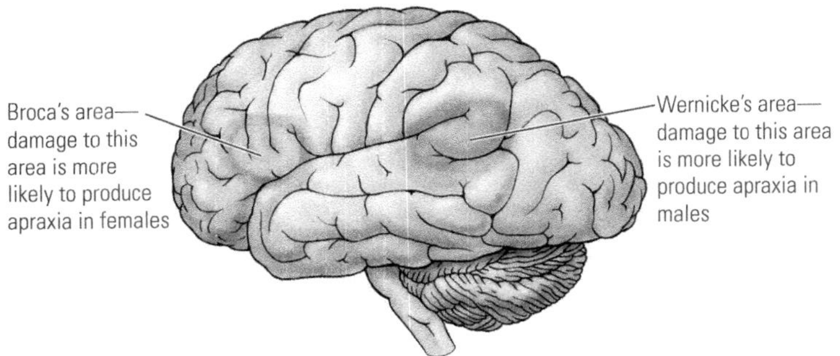

Crossword Puzzle

[1]N	E	U	R	O	S	C	[2]I	E	N	C	E				[3]E									
							N								X						[4]C			
			[5]F	U	N	C	T	I	O	N	A	L		[6]A	T	T	E	N	T	I	O	N		
			M				E								I						R			
			R				L		[7]M	[8]F	B		[9]S	Y	N	T	A	X			T			
[10]S	[11]P	L	I	T			L			A			Y		C						E			
	L						I			[12]C	O	G	N	I	T	I	O	N			X			
	A						[13]G	[14]A	M	E			E		I				[15]C					
	N					[16]M	E	G			[17]R		S		O		[18]A		O					
	N						[19]N	E	G	L	E	C	T		N		P		N			[20]M		
	I						C				S		H				R		V			A		
	N						E		[21]C		O		E				A		E			G		
	[22]G	A	G						O		N		S		[23]R		X		R			N		
									M		A		[24]I	[25]M	A	G	I	N	G			E		
							[26]B	R	A	I	N		[27]A	R	T		A		E			T		
											C			I					N			I		
								[28]R	E	T	E	S	T		[29]C	O	N	S	T	R	U	C	T	S

Chapter 16: What Happens When the Brain Misbehaves?

Multiple-Choice Questions

1. C (p. 568)
2. E (p. 572)
3. C (p. 574)
4. B (p. 575)
5. E (p. 575)
6. B (p. 578)
7. E (p. 578)
8. A (p. 579)
9. C (p. 584)
10. C (p. 584)
11. B (p. 586)
12. B (p. 587)
13. E (p. 587)
14. B (p. 588)
15. B (p. 591)
16. D (p. 593)
17. E (p. 593)
18. C (p. 597)
19. A (p. 600)
20. A (p. 602)

Short-Answer Questions

1. (p. 568)
2. (p. 569)
3. (p. 577)
4. (p. 577)
5. (p. 586)
6. (p. 592)
7. (p. 597)
8. (p. 600)
9. (p. 602)
10. (p. 603)

Matching Questions

1. _B_ To date, has produced only modest improvements in patients
   _A_ Patients often relapse when treatment is "turned off"
   _C_ Tends to produce significant memory loss
   _D_ Potential for significant long and short-term side effects

2. _DSM_ American Psychiatric Association Publication
   PKU Genetic disorder treated with dietary restrictions
   DBS Neurosurgical approach to treating Parkinson's disease
   TMS Most recently developed seizure therapy method
   MRS Can identify changes in specific markers of neural function

3. _+_ Tremor at rest
   _−_ Akinesia
   _+_ Oculogyric crisis
   _+_ Muscular rigidity
   _−_ Disorders of posture
   _−_ Disorders of locomotion

4. _C_ Epilepsy
   _A_ Parkinson's disease
   _D_ Alzheimer's disease
   _B_ Stroke

5. _C_ Posttraumatic stress disorder
   _A_ Schizophrenia
   _B_ Bipolar disorder
   _C_ Obsessive-compulsive disorder
   _B_ Mania

Diagrams

1.

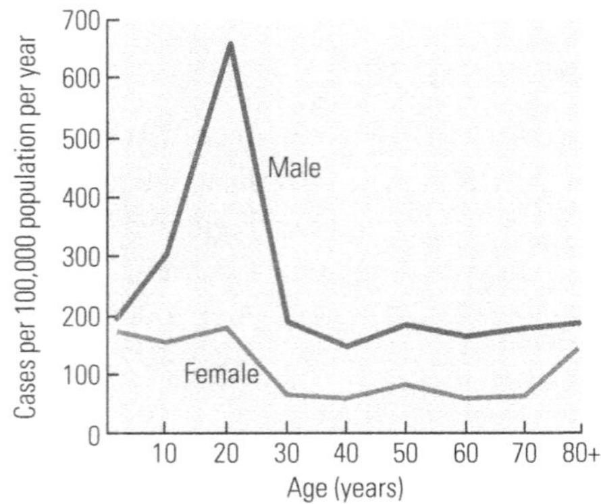

2.

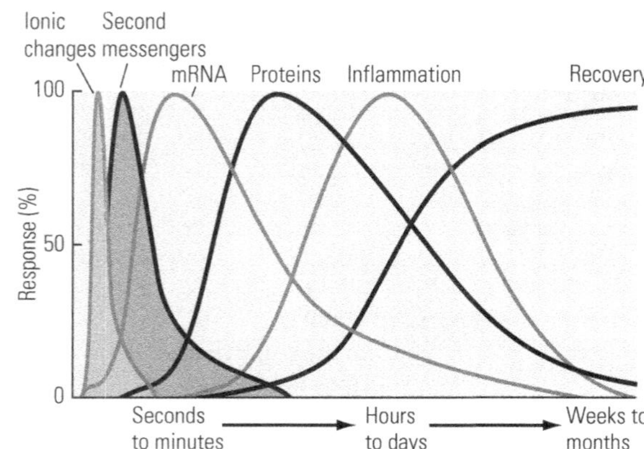

3.

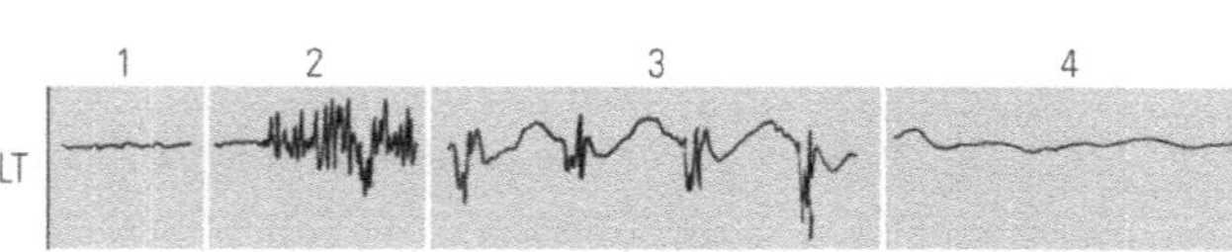

4.

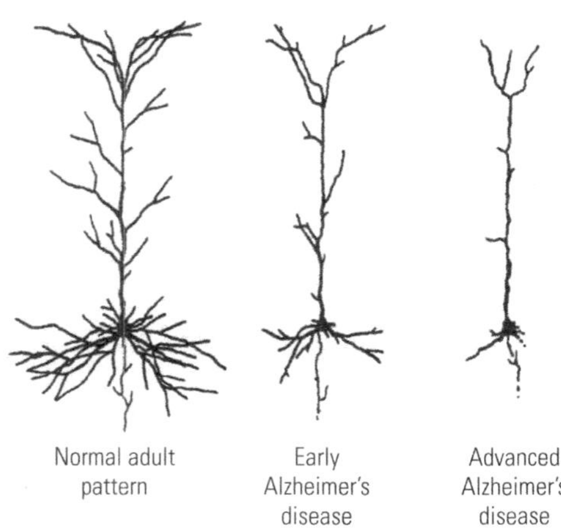

4.

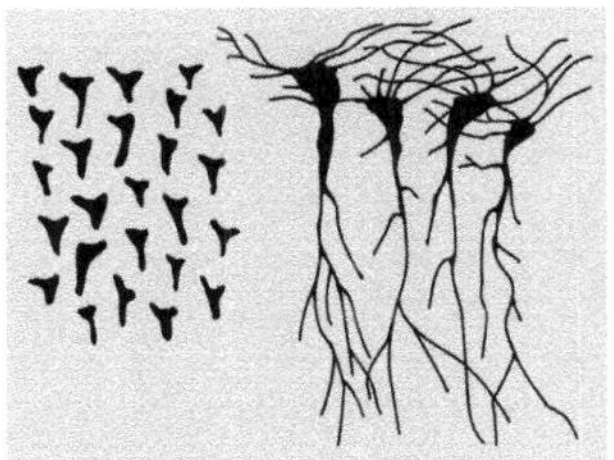

Organized (normal) pyramidal neurons

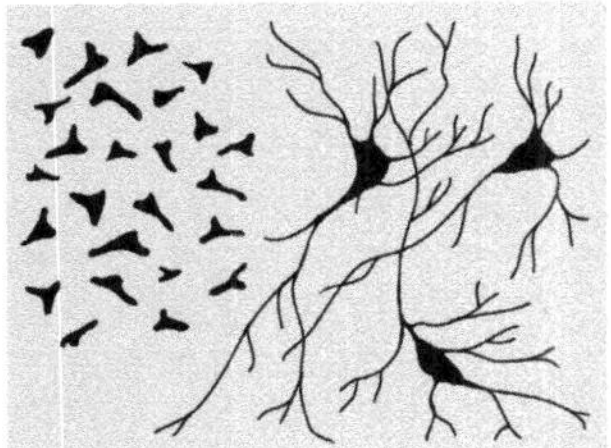

Disorganized (schizophrenic) pyramidal neurons

Crossword Puzzle

1 D	E	S	E	N	S	I	T	I	Z	A	T	2 I	O	N							
E												I									
3 G	E	N	E	R	4 A	L	I	Z	E	D											
E					L							5 P		6 B							
N		7 E			Z				8 B			E		O							
E		C			H				I			T		D		9 S					
R		10 T	Y	P	E				11 P	O	S	I	T	I	V	E			12 M		
A					I				O			T		E		I			R		13 B
T					M				L					S		Z		14 M	S		O
I					E			15 M	A	N	U	16 A	L		17 A	U	R	A			D
V				18 T	R	E	M	O	R			N				R		19 L	E	W	Y
E			20 D	M	S			O			21 A	X	I	S		E					
				S				D				I				S					
											22 D	E	E	23 P							
												T		K							
												Y		U